Operational Control of Coagulation and Filtration Processes

AWWA MANUAL M37

Second Edition

American Water Works Association

Science and Technology

AWWA unites the entire water community by developing and distributing authoritative scientific and technological knowledge. Through its members, AWWA develops industry standards for products and processes that advance public health and safety. AWWA also provides quality improvement programs for water and wastewater utilities.

MANUAL OF WATER SUPPLY PRACTICES—M37, Second Edition

Operational Control of Coagulation and Filtration Processes

Project manager and copy editor: David Talley
Production editor: Carol Magin

Library of Congress Cataloging-in-Publication Data

Operational control of coagulation and filtration processes.
    p. cm. -- (AWWA manual ; M37)
  Includes bibliographical references and index.
  ISBN 1-58321-055-5
    1. Water--Purification--Coagulation. 2. Water--Purification--Disinfection. I. AWWA
Research Foundation. II. Series.

TD491 .A49 no. M37 2000
[TD455]
628'.1 s--dc21
[628.1'64]   99-088034

Printed in the United States of America
American Water Works Association
6666 West Quincy Avenue
Denver, CO 80235

ISBN 1-58321-055-5

Printed on recycled paper

Contents

Figures

Tables

This page intentionally blank.

Foreword

The first edition of AWWA Manual M37 was prepared by the Coagulation and Filtration Committee of the AWWA Water Quality Division under the direction of David A. Cornwell, who served as overall coordinator and technical editor.

This second, revised edition of the manual was also prepared by the Coagulation and Filtration Committee of the AWWA Water Quality Division under the direction of David J. Hiltebrand, with special assistance from Peter Pommerenk in reviewing the manual for content and consistency. Matt Alvarez and Gary Schafran provided additional reviews and recommendations. The members of the M37 subcommittee included:

David J. Hiltebrand, Chair, AH Environmental Consultants, Inc.,
    Newport News, Va.
Howard J. Dunn, Bridgeport Hydraulic Company, Bridgeport, Conn.
C. Michael Elliott, Stearns & Wheeler, Cazenovia, N.Y.
Christopher B. Lind, General Chemical Corporation, Syracuse, N.Y.
Nancy E. McTigue, Environmental Engineering & Technology, Newport
    News, Va.

Peter Pommerenk served as a guest author for the revisions to Chapter 1. Steven K. Dentel, University of Delaware, and Paul Zielinski, Pennsylvania American Water Company, served as information resources and guest authors for the revised Chapter 2.

Much of the information from the first edition of this manual has been retained, so it is important to recognize the original committee members who developed that material. They include:

David A. Cornwell, Chair, Environmental Engineering & Technology, Newport
    News, Va.
Charles F. Anderson Jr., City of Arlington, Arlington, Texas
Robert W. Bailey, CH2M Hill Consulting Engineers, Orlando, Fla.
Keith E. Carns, Carns, Perkins & Associates, Pinole, Calif.
Howard J. Dunn, Regional Water Authority, New Haven, Conn.
C. Michael Elliott, Stearns & Wheeler, Cazenovia, N.Y.
Gilbert Faustel, Albany, N.Y.
Charles A. (Skip) Griffin, Jr., Camp Dresser & McKee, Carlsbad, Calif.
David J. Hiltebrand, Malcolm Pirnie, Inc., Yorktown, Va.
Greg J. Kirmeyer, Economic & Engineering Services, Bellevue, Wash.
Allen L. Lange, Contra Costa Water District, Concord, Calif.
Raymond D. Letterman, Syracuse University, Department of Civil and
    Environmental Engineering, Syracuse, N.Y.
Christopher B. Lind, General Chemical Corporation, Syracuse, N.Y.
Gary S. Logsdon, Black & Veatch Engineers, Cincinnati, Ohio
Donald G. McBride, City of Los Angeles, Department of Water & Power,
    Los Angeles, Calif.
Nancy E. McTigue, Environmental Engineering & Technology, Newport
    News, Va.
William E. Neuman, American Water Works Service Company, Voorhees, N.J.

Naeem Qureshi, Progressive Consulting Engineers, Minneapolis, Minn.

Larry Scanlan, Central Utah Water Conservancy District, Orem, Utah

Robert Hoehn, Division Liaison, Virginia Polytechnic Institute and State University, Blacksburg, Va.

For the first edition:

Chapter 1 was chaired by Skip Griffin with co-authors Dave Hiltebrand and Chris Lind. Significant assistance was given by Ersin Kasirga of the staff of Environmental Engineering & Technology, Inc.

Chapter 2 was chaired by Michael Elliott, with William Neuman as co-author. Steven Dentel, University of Delaware, Newark, served as a resource and guest author for this chapter.

Chapter 3 was chaired by Nancy McTigue, with Howard Dunn as co-author. Ken Burman, City of Tulsa, Tulsa, Okla., was a guest author for this chapter.

Chapter 4 was prepared by Ray Letterman.

Chapter 5 was chaired by Bob Bailey, with Bill Bellamy of CH2M Hill, Denver, Colo., as guest author.

Gary Logsdon served as a reviewer.

Introduction

Early water treatment was performed to improve appearance or taste. With the discovery that diseases such as cholera and typhoid fever were transmitted by ingestion of water contaminated with particulate germs, people came to recognize that taste and smell alone are not accurate indicators of the acceptability of water. As a result, additional treatment technologies beyond the commonly used sedimentation basins were needed. Experimental work in Louisville, Ky., Cincinnati, Ohio, and Pittsburgh, Pa., brought about the development of the rapid sand filters, which were shown to significantly reduce both turbidity and bacteria in water. These filters required addition of a coagulant to allow the coarse sand to remove colloidal materials. The coagulant was usually added to the raw water as it entered the sedimentation basin. After a few hours of contact in the basin, the water was sent to rapid sand filters. To control the coagulation process, treatment performance was often "eyeballed" by observing the floc formation in the sedimentation basin and adjusting the chemical dose to maintain a "beautiful-looking" basin.

Another milestone in drinking water treatment was the use of chlorine as a disinfectant. Introduced in the United States at the beginning of the 20th century, this practice led to a dramatic reduction in the occurrence of waterborne diseases. This combination of coagulation, sedimentation, filtration, and disinfection constituted an early multiple-barrier approach to microbial control in drinking water technology.

Traditionally, conventional treatment has focused on turbidity as the primary indicator of process performance and efficiency. Disinfection has focused on the control of coliform organisms, generally by adding chlorine somewhere within the treatment process and maintaining a disinfectant residual through the treatment plant and within the distribution system. This approach to the treatment of surface waters, though somewhat simplistic by today's standards, was sufficient to help eliminate major outbreaks of cholera and typhoid.

While these classic examples of waterborne diseases have been controlled, concern remains about illnesses transmitted by newly recognized agents such as viruses, certain bacteria, and protozoans such as *Giardia* and *Cryptosporidium*. Additional concerns deal with the chemical compounds, known as disinfection by-products (DBPs), that are formed whenever an oxidant such as chlorine is added to a water that contains natural organic material (NOM). As a result of these concerns, new and pending drinking water regulations have significantly shifted the focus of conventional surface water treatment:

- The concept of microbial control has been expanded to include not only coliform organisms but also other microorganisms that are more difficult to control, such as *Giardia*, *Cryptosporidium*, and viruses.

- The role of turbidity as an indicator of particle removal has been expanded to function as a surrogate indicator of microbial removal and interference with disinfection.

- The Surface Water Treatment Rule (SWTR) formally established the multiple-barrier approach to microbial control, including steps for removal of contaminants by coagulation/filtration, primary disinfection as defined by the *CT* concept, and maintenance of microbial control through the distribution system.

- The enhanced coagulation requirements of the new Disinfectants/Disinfection By-products Rule (D/DBPR) sets targets for the removal of NOM that is usually present in surface water supplies.

In many cases, the traditional approach to conventional treatment is no longer sufficient to meet the new requirements. Water treatment is becoming more sophisticated and costly, creating a need to look at old processes in new ways along with the continuing need to look at new processes. One of the more effective methods of meeting the new requirements is through a systems approach recognizing that all unit processes are interrelated, so what impacts one will also impact the others. Analysis under this systems approach shows that most conventional treatment processes exist to remove "things" from the water: NOM, color, particles, microorganisms, iron, manganese, objectionable tastes and odors, etc. To remove these target constituents, the conventional process designs set two goals:

- Create particles by converting organic and/or inorganic material from the dissolved phase into the particulate phase

- Condition particles for subsequent removal by either clarification or filtration

Much of the effort to "optimize" conventional treatment must focus on pretreatment chemistry as the single most important factor affecting treatment plant performance. This principle is based on the very real fact that if the pretreatment chemistry is wrong, none of the other downstream processes will work.

In order to optimize pretreatment processes, designers and operators must understand:

- What is in the raw water that controls the treatment processes

- Treatment criteria needed to produce the best possible water

NOM is formed in water from the biodegradation and/or transformation of plant and animal material in the watershed. This biogenic material, consisting of complex organic molecules, stabilizes particulate matter through adsorption. It is the main precursor for organic DBPs. Its binding capacity enables mobilization of metals and synthetic organic compounds, and it also serves as a substrate for microbial growth in distribution systems.

In addition, this NOM provides the most significant contribution of negative charges to be neutralized by positively charged coagulants. As a result, the nature and concentration of the NOM in a water most often controls the selection of coagulant type and dose.

To satisfy the enhanced coagulation requirements of the D/DBPR, many surface water plants are trying to optimize NOM removal. Often, this effort involves high dosages of alum or iron salts combined with coagulation at low pH. These practices were initiated without recognition that alum consumes 0.5 mg/L of alkalinity as $CaCO_3$ for each mg/L of alum; each mg/L of ferric chloride consumes 0.93 mg/L of alkalinity as $CaCO_3$. Increasing the alum or ferric dose beyond the available alkalinity without the addition of lime or caustic allows unreacted coagulant to pass through the treatment plant, possibly leading to precipitation when the finished water pH adjustments are made.

In addition, the conditions that provide for optimum NOM removal may not favor effective particle removal. Since particle removal is an integral part of the multiple-barrier approach to microbial control, it must not be sacrificed or compromised in any way. As a result, treatment may need to provide for supplemental polymer addition when practicing enhanced coagulation. To ensure effective particle removal, for example, the pH of coagulation can have a significant impact on the downstream processes. Traditionally, a pH range of 6.5 to 7.2 was thought to be optimum for particle removal with alum coagulation. With the move toward optimizing NOM removal, many utilities are evaluating coagulation at pH 5.0 to 5.5. While low pH conditions favor the removal of NOM, particle removal may suffer. Also, not all treatment plants are constructed of materials that can withstand acidic pH ranges for long periods of time.

A second area of concern is beyond the scope of this manual: the impact of these process changes on water quality in the distribution system. With the implementation of the Lead and Copper Rule, the water industry discovered that significant changes in the chemistry of the water entering or stored within the distribution system could destabilize any protective film created by corrosion control treatments and reduce effectiveness of those treatments. Since then, this recognition of the effect of chemistry changes on the stability of materials attached to the walls of the distribution system piping has extended to include changes created by new and modified treatment plants and by blending new supplies.

Once the chemistry has been optimized, process control decisions focus on the physical aspects of pretreatment, namely the mixing and clarification processes. Recognizing that the chemical reactions associated with coagulation occur within seconds, the rapid mix step must accomplish its purpose of dispersing treatment chemicals in the raw water as quickly and efficiently as possible. For any given water supply, empirical trials must determine the mixing intensities and detention times required.

Within the treatment process, flocculation changes the size distribution of the particles that have been created or conditioned. A large number of small particles are transformed into a smaller number of larger particles. Traditionally, the objective of flocculation has been to produce particles large enough and dense enough to settle in the clarifier. Process designs seek to produce these larger particles by fluid motion. The primary problems associated with flocculation tend to be short circuiting or, in processes using turbine impellers, localized shearing due to tip velocities greater than 8 fps.

Clarification was once viewed as a simple bump in the line where the water slowed down sufficiently to allow large, heavy particles to settle by gravity. This step is now recognized as playing a more active role within the treatment plant. Professor Charles R. O'Melia (1998) has emphasized these points:

- In addition to gravity settling, coagulation occurs in all settling facilities receiving particles that can attach when they come into contact. These contacts occur among the particles by diffusion and differential settling.

- A sludge blanket clarifier combines the processes of sedimentation and filtration where particles entering the basin can collide with particles previously retained in the sludge blanket.

- These changes in particle concentrations, size distributions, and surface characteristics have a profound impact on subsequent filtration.

Increasingly sophisticated instrumentation and analytical techniques, such as particle counters and microbial particulate analysis, have demonstrated a need for continuing improvement. Even under optimum conditions, granular media filters can pass large quantities of particles, including algae and other microorganisms, into the finished water, compromising efforts at water quality control in the distribution system. Where surface chemistry may not be conducive to conventional coagulation/filtration, adequate removal of these constituents may require an alternative clarification process such as dissolved air flotation.

The performance of granular media filters reflects two factors:

- Source water quality

- Changes induced by added chemicals

Filtration is not performed by granular media particles alone. It is not just a physical straining process like that in a coffee filter. The particles that attach to the media perform the filtration. Filtration is a physical and chemical process in which the effectiveness of the particle removal is determined by several variables:

- Type of filter media (size, depth, material)

- Water chemistry

- Surface chemistry of the particles

- Surface chemistry of the media

Particles must be transported to the surfaces of the filter media, and they must attach there, for removal to occur. Both hydrodynamics and chemistry are important determinants of success. Design criteria most often specify filtration rate, media size, and bed depth. These variables are pertinent to filter performance, but they are not sufficient alone to assure filter effectiveness. In fact, they usually have less impact on the process than several other properties of the system.

Pretreatment chemistry is the most important factor affecting particle removal in granular media filters. Without proper control of the solution chemistry of the water to be treated, the surface chemistry of the particles to be removed, and the surface chemistry of the media used to accomplish removal, efficient particle removal will not occur, regardless of the filtration rate, bed depth, and media size.

Therefore, when a problem is recognized with quality of filtered water, and investigation confirms that the filters have not physically failed, a good place to start further analysis is either the water chemistry or the surface chemistry of the particles or the media. Options for altering the water chemistry and/or particle surface chemistry include changes to

- Coagulant type and dose

- pH of coagulation

- Polymer type and dose

- Oxidant type and dose

Options for altering the surface chemistry of the media include but are not limited to the use of a filter aid.

Head loss development is a function of the concentration of particles and the particle size distribution. Therefore, problems with excessive head loss buildup tend to be due to mechanical deficiencies such as inadequate flocculation.

As previously indicated, the key to meeting challenges set by new regulations will often be maximizing the removal of NOM while ensuring adequate microbial control, i.e., particle removal and disinfection. Optimizing treatment processes to reach these goals may involve setting coagulant doses to achieve NOM removal, reducing coagulation pH, improving mixing conditions, or using alternative oxidants. Consequently, water treatment operators need effective tools for quick and accurate assessments of treatment performance and evaluations of alternatives.

This manual discusses the commonly available methods by which operators can optimize coagulation and filtration processes and maintain day-to-day control. Throughout all the innovations and changes within the water treatment field, the traditional jar test method, presented in chapter 1, remains the predominant method for evaluating coagulation. Therefore, the discussion of this technique occupies the most space in this manual. Streaming current detectors (chapter 2) and electrophoretic mobility measurements (chapter 4) are electrokinetic methods for assessing particle/colloid properties relevant for conventional water treatment. Use of these methods may assist in optimizing coagulant doses in bench-scale studies and during plant operation. Chapter 3 provides information about particle sizing and counting. This method is an excellent tool for studying the impacts of different mixing regimes on particle formation and evaluating the performance of clarification and filtration units in treatment plants. Chapter 5 deals exclusively with the optimization of filtration through the use of pilot filters.

REFERENCE

O'Melia, C.R. 1998. Coagulation and Sedimentation in Lakes, Reservoirs, and Water Treatment Plants. *Wat. Sci. Technol.* 37:2:129–135.

This page intentionally blank.

Chapter **1**

Jar Testing

GENERAL THEORY

The jar test is recognized throughout the water industry as the most valuable and most commonly used tool for realistically simulating coagulation control at a full-scale treatment plant. This test is routinely performed by water treatment plant operators as well as consultants and researchers. It basically involves duplicating, sequentially in a single vessel, conventional treatment steps that occur simultaneously at different locations in the plant. Jar testing may be used to evaluate the effects of changes in chemical doses and points of application, choosing alternative coagulants, adding polymeric coagulant aids, implementing alternative preoxidation strategies, varying mixing intensities and times, and changing overflow rates on the removal of particles, natural organic matter, or other water quality parameters of concern.

Successful simulation of treatment plant conditions requires knowledge of the hydraulic characteristics of the treatment steps with initial mixing (also called *rapid* or *flash mixing*), flocculation, and clarification, as well as translation into a batch-testing protocol. The key parameters include:

- Velocity gradient in the mixing basin
- Effective retention time in the mixing basin
- Velocity gradient in the flocculation basin
- Effective retention time in the flocculation basin
- Surface loading rate of the sedimentation basin

Velocity Gradient

The concept of the velocity gradient is one of the most important ideas to be understood when conducting jar tests. The intensity of mixing is generally measured by the velocity gradient. A higher number indicates more intense mixing. Velocity gradient is often expressed as G with units of s^{-1} (seconds to the minus 1 power). The velocity gradient is calculated using the energy dissipation rate in the fluid, or it can be interpolated from

1

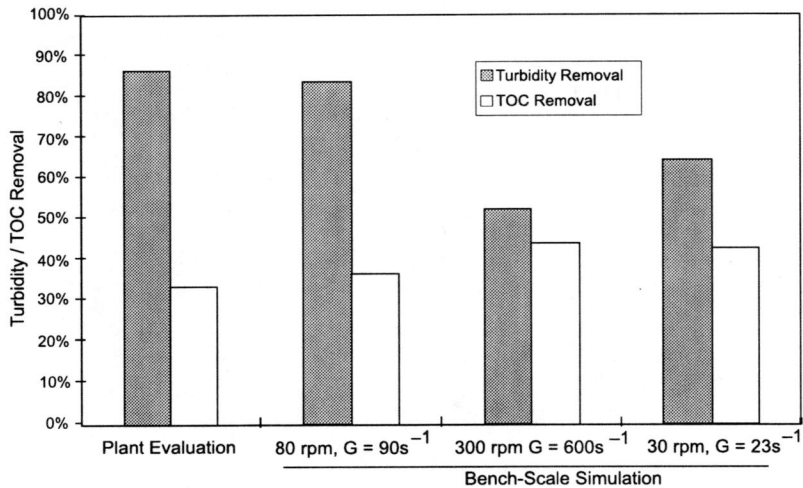

Source: AH Environmental Consultants, Inc. (1998).

Figure 1-1   Use of jar tests to determine optimum flash mix conditions

calibration curves. In order to receive comparable results, mixing intensities—and therefore velocity gradients—during jar tests should correspond to those in the treatment plant. Existing empirical relationships can be used to determine the appropriate impeller speed for particular jar test equipment to achieve a given velocity gradient.

Initial Mixing

The purpose of the initial mixing phase in a jar test is to completely disperse the primary coagulant into the raw water flow stream in a fast and uniform manner. This mixing makes the coagulation process as effective as possible. The term *initial mixing* is intended to apply to the point at which the primary coagulant is added to the water. Hydrolysis takes place almost instantaneously during the addition of the coagulant. Complete floc development, on the other hand, may take several minutes. Two commonly used terms for initial mixing—*rapid mixing* and *flash mixing*—are often used interchangeably.

When conducting a jar test, the time the mixer runs at rapid mix should be equivalent to the effective retention time of the mixing chamber in the treatment plant. If possible, G values should match full-scale conditions, as well. Figure 1-1 illustrates how turbidity and TOC removal in jar testing can vary with mixing speed. Apparently, for the water supply in the figure, a G value of 90 s^{-1} simulates the full-scale conditions best.

Flocculation

Flocculation refers to a gentle mixing process that occurs in the flocculation basin. A primary objective of the water treatment process is to destabilize and condition the particulate matter in water chemically and physically to encourage settling and/or filtration. Particle destabilization is caused by coagulant addition. The subsequent increase in the size of particles results in the development of floc through the process called *flocculation*.

A properly designed and performed set of jar tests can give the operator a tremendous advantage in determining the correct coagulant and dose for developing the optimum floc. The word *floc* originates from the Latin root word *floccus,* and it describes a grouping of solid particles with a woolly appearance (Scott and Smith,

1980). The size and strength of the floc developed determines the efficiency of the solids removal process. Floc size can be easily observed in the jar test and related to the settling velocity. Floc strength or toughness may be defined as resistance to fragmentation by shear forces induced by hydraulic velocity gradients (Hannah, Cohen, and Roebeck, 1985). This characteristic cannot be quantified with the jar test.

Appropriate floc characteristics vary depending on the type of treatment process. A typical conventional surface water treatment process utilizing sedimentation followed by filtration benefits from a large, heavy floc to facilitate settling. A common misconception is that increasing the chemical dose increases the settling efficiency of the floc particles, because the larger flocs look great in the basins. These larger particles are quite often simply a chemical floc, such as aluminum hydroxide precipitate, which has a near-neutral buoyancy and very little strength but a great appearance. In reality, this dose increase is a waste of chemicals and money that creates more residual solids than is necessary.

If the floc is too weak to withstand the shear forces in the sedimentation and/or filtration stages of a conventional process, the floc will eventually penetrate the filters. The carry-through floc will degrade finished water quality and create problems in the distribution system.

A direct filtration process requires a floc strong enough and small enough to facilitate good filter performance without breaking up and causing premature floc breakthrough. Flocs that are too large may quickly clog filters.

In jar testing, the retention time and G values of the flocculation basin should correspond to those in the water treatment plant. This condition may require varying the mixing speeds, if the plant performs tapered flocculation in different compartments.

Sedimentation

Sedimentation basins remove particles by gravity. The particles to be removed may be either naturally occurring materials or the results of the coagulation/flocculation process.

The surface loading or overflow rate is the most important parameter for sedimentation. It is determined by dividing the basin flow rate by the surface area, commonly expressed in gallons per day per square foot (gpd/ft^2). A basic unit conversion reveals that the surface loading corresponds to velocity:

$$\text{Surface loading rate} = \frac{\text{Flow rate}}{\text{Surface area}} = \frac{\text{Volume/Time}}{\text{Area}} = \frac{\text{Length}}{\text{Time}} \qquad \text{(Eq 1-1)}$$

or

$$\text{gpd/ft}^2 \times 3{,}785 \text{ cm}^3\text{/gal} \times 1 \text{ day/1,440 min} \times 1 \text{ ft}^2\text{/929 cm}^2 \rightarrow \text{cm/min} \qquad \text{(Eq 1-2)}$$

which is equivalent to a settling velocity or the rate at which the particles settle. To calculate the settling velocity (stated in cm/min) from the plant flow rate and the total sedimentation basin area the following conversion factors can be used:

Plant Flow Rate / Surface Area	Multiply by	Settling Velocity
mgd/ft^2	2,829.56	cm/min
gpm/ft^2	4.0746	cm/min

The following table lists surface loading rates and corresponding settling velocities:

Surface Loading Rate		Settling Velocity
gpd/ft²	m/h	cm/min
180	0.3	0.5
360	0.6	1.0
720	1.2	2.0
1,440	2.4	4.0
3,600	6.0	10.0

Note: 1 gpm/ft² (2.4 m/h) is equivalent to a settling velocity of 4 cm/min.

Theoretically, a sedimentation tank will remove all particles that exceed this critical velocity for a given overflow rate. This concept is most important for simulating the sedimentation process in a jar test. The retention time of the full-scale settling basin is neither correct nor practical in the jar test. In order to obtain useful results, the test's surface loading, which corresponds to a settling velocity, must match that of the process, as indicated earlier. In jar testing, this match is accomplished by collecting a sample from the jar at a set depth below the water surface at a given time.

For example, a typical sedimentation basin is designed for an overflow rate of 700 gpd/ft². This value corresponds to a particle settling velocity of 2 cm/min. If the sampling port of the jar is 10 cm below the water surface, then all particles that pass the sampling port within 5 min (10 cm divided by 2 cm/min) would be removed. Therefore, samples should be collected 5 min after the flocculation period to simulate the performance of the sedimentation basin. If the jar were sampled on the basis of the settling basin's retention time (e.g., 120 min), the performance of the basin would be grossly overestimated.

Figure 1-2 illustrates a minimal effect of polymer addition for samples taken after 10 min. When the same plant is operating at a maximum surface loading rate of 1,800 gpd/ft², which corresponds to a sampling time of 2 min, the effect is more pronounced.

For jars in which the sample port is 10 cm below the surface, the following table lists settling velocities and corresponding sampling times:

Settling Velocity	Sampling Time
cm/min	min
0.5	20
1.0	10
2.0	5
4.0	2.5
10.0	1

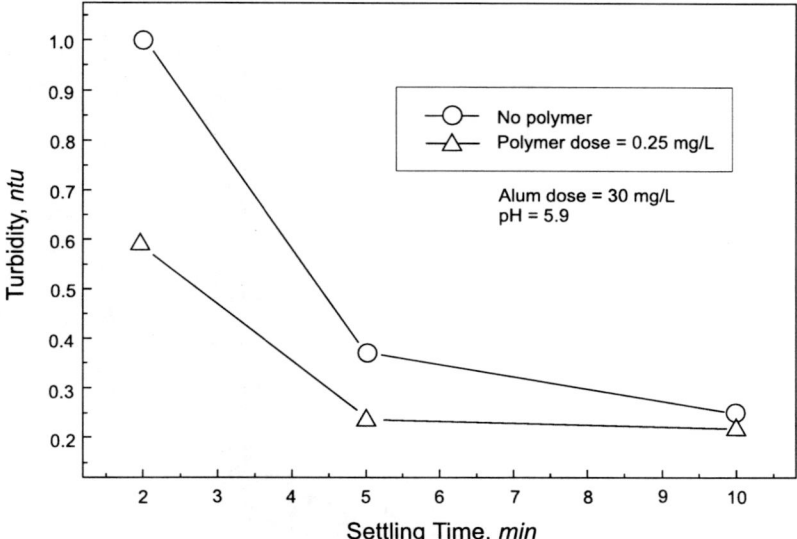

Source: AH Environmental Consultants, Inc. (1997).

Figure 1-2 Example use of jar tests: Settled turbidity versus settling time

Point of Application

If the purpose of a jar test is to optimize conditions for the system under consideration, chemicals should be added in the same order as in the water treatment plant. The same chemicals can be added at different times in an attempt to validate existing practices or to improve the quality of water produced. For example, a jar test might evaluate potential gains from polymer addition along with coagulant addition during a second rapid mix stage in a flocculation compartment.

Most operators work with predetermined points of application because of the constraints of the constructed facilities. Jar tests might show improvements through adding chemicals at different locations, perhaps justifying modifications to the existing plant. If the treatment plant has not yet been built, predesign verification through piloting and bench-scale studies, i.e., jar testing, provides data not available otherwise. Raw waters vary in their chemical, physical, and biological characteristics; the only sure way to determine the most efficient point of coagulant application short of full-scale plant trials is through a pilot-plant study or jar testing. Well-established principles confirm that varying application points may be necessary to optimize a treatment process. Singley (1981) reported on one case in which addition of lime prior to the coagulant resulted in much better settling. In studies conducted in Arlington, Texas (Environmental Science & Engineering, 1981), several sets of jar tests indicated that optimum full-scale results would result from adding polymer at the exit of the rapid mix just before water entered the first stage of flocculation. The study evaluated polymer addition prior to alum, with alum, after the first stage of flocculation, and at the end of the rapid mix. Figure 1-3 graphically illustrates the results of the study.

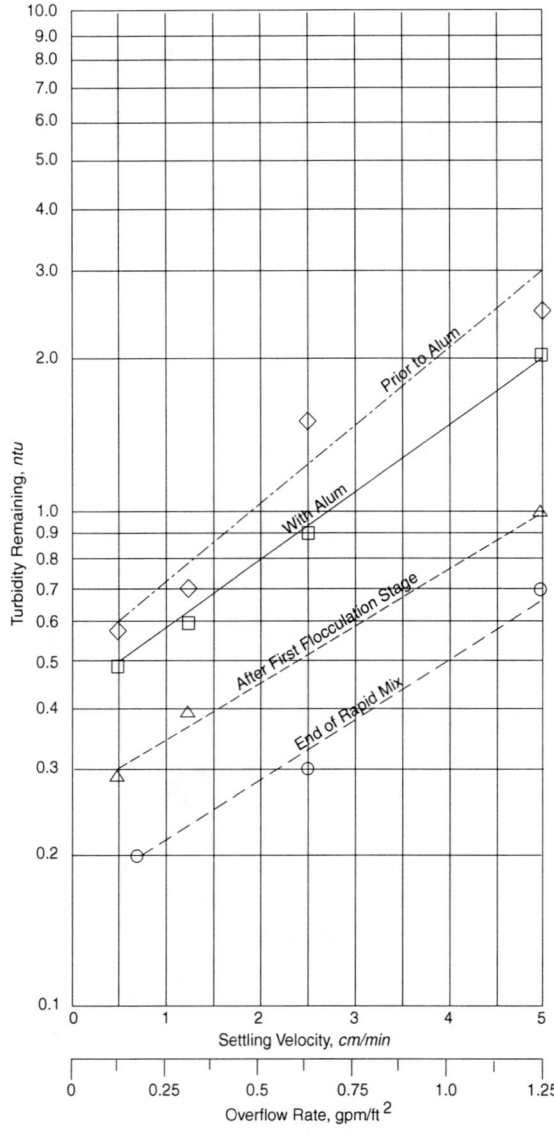

Source: Environmental Science & Engineering, Inc. (1981).

Figure 1-3   Example jar test: Point of chemical addition

PREPARING FOR A JAR TEST

The formulation of a realistic and useful jar test protocol depends on several major steps. These steps require time and effort on the part of the operator, but the gain will be well worth the effort. A well-planned and organized approach will save time in the long run by reducing guesswork and identifying the critical resources and information needed to ensure that a jar test yields useful data.

Defining Study Goals

Jar tests can be conducted for a number of reasons. The most common use of the jar test is for day-to-day process control, but it may also be used to demonstrate compliance or design new facilities. To plan a jar test for an existing facility, study goals may include evaluating the effects of process changes on treated water quality, as follows:

- Changes in chemical doses, for example, determining the point of diminishing returns (PODR) or the optimum coagulation pH

- Alternative chemical choices, for example, coagulants (ferric chloride, alum, polyaluminum chloride, etc.), pH-controls (lime versus caustic soda), or preoxidants (potassium permanganate, ozone, chlorine dioxide, etc.)

- Additional chemical choices, for example, organic polymers for enhanced solids–liquid separation

- Physical modifications, such as varying mixing intensities and points of chemical application, implementing tapered flocculation, or installing baffles to increase retention times

Similarly, possible procedures for new physical facilities can be evaluated. Water quality parameters that are commonly used to assess treatment performance include:

- Turbidity and color removal

- Dissolved organic carbon concentration or UV-254 absorbance

- Disinfectant by-product formation

- Iron and manganese removal

- Taste and odor of the treated water

- Sludge characteristics

Required Information

The second step in preparing a useful jar test protocol is to gather the necessary information. The list includes data about the physical characteristics of the plant, applications of chemicals, and current treatment performance.

To make valid comparisons between jar test data and actual plant performance data, an operator needs to know the key parameters of the existing or future plant equipment. These factors should include velocity gradients or mixing intensities and detention times for mixing, flocculation, and channels or pipes, as well as overflow rates for sedimentation basins.

Velocity gradient. The design documents or operations and maintenance manual for a water treatment plant should provide data for actual in-plant velocity gradients. Figure 1-4 shows an example of a chart relating velocity gradient to mixer speed and water temperature. The accuracy of this information is very important for reliable results. If the information is not readily available, consult the article by Cornwell and Bishop (1983) for axial or turbine mixers. Fair, Geyer, and Okun (1968) cover paddle flocculators.

Detention times. The theoretical detention or residence time is defined by the following equation:

$$\text{Detention time} = \frac{\text{Tank volume}}{\text{Flow rate}} \qquad \text{(Eq 1-3)}$$

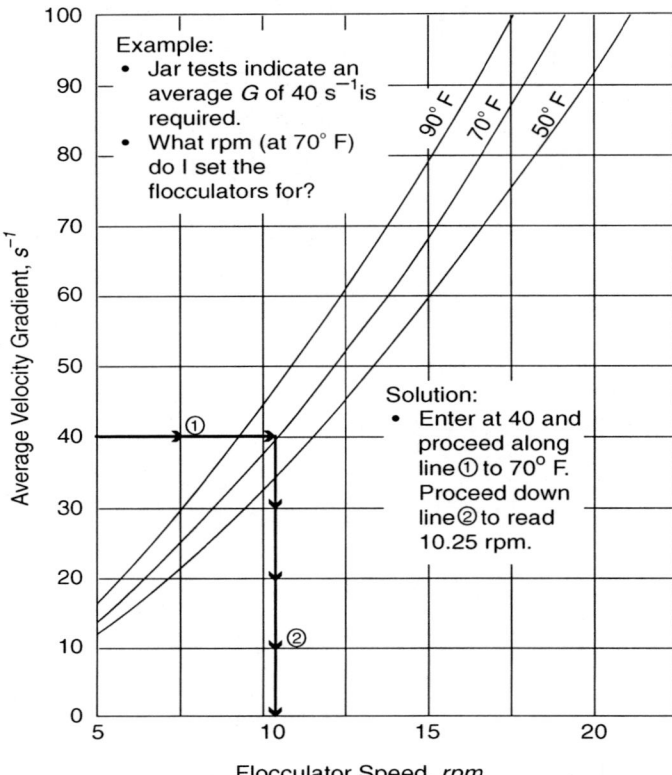

Visual observation of stage 2 or 3 mechanisms

Source: Environmental Science & Engineering, Inc. (1981).

Figure 1-4   Example use of graph for determining velocity gradient for jar test based on full scale

For example, the detention time of a tank with the dimensions 7 ft (2.14 m) × 6 ft (1.83 m) × 10 ft (3.06 m) (depth × width × length) operated at a flow rate of 105 gpm (23.9 m³/h) would be

$$\frac{7\ \text{ft} \times 6\ \text{ft} \times 10\ \text{ft}}{105\ \text{gal/min}} \times 7.48\ \text{gal/ft}^3 = 30\ \text{min} \qquad \text{(Eq 1-4)}$$

$$\frac{2.14\ \text{m} \times 1.83\ \text{m} \times 3.06\ \text{m}}{23.9\ \text{m}^3/\text{h}} \times 60\ \text{min/h} = 30\ \text{min} \qquad \text{(Eq 1-5)}$$

In this way, theoretical detention times can be readily calculated from fixed tank dimensions and the flow rate. However, actual detention times through a plant's mixing basins may be less than one-half of the theoretical detention times, perhaps due to short-circuiting in the process units. Tracer studies may be necessary to determine the actual detention time.

It should also be noted that the procedure defined earlier for jar tests is not always used just to find the equivalent mixing time. Duration of mixing also affects mixing conditions. Standard practices for jar testing include collection of turbidity samples at various times (min) after stirring has stopped in order to develop settling velocity curves, but the contact time in the full-scale basin may be 2 to 4 h. The result is that for a given settling velocity, the time of sedimentation in the jar test is about 1 one-hundredth that in the conventional plant. Therefore, reactions that may be occurring in the basin, such as formation of disinfectant by-products, may not be simulated in the jars. If the bench testing procedure is used to determine a contact-time-dependent parameter, such as formation of THMs during the treatment process, it may be necessary to hold the jars for that period of time.

Sedimentation basin overflow rate. The sedimentation basin overflow rate, or surface loading, determines the length of time that a water is allowed to settle in a jar before a sample is taken. This time is based on the relationship of surface loading and settling velocity, as previously discussed. If the flow rate and the surface area of the sedimentation basin(s) are known, the overflow rate can be calculated by dividing the flow rate by the surface area.

Chemicals and points of application. Another planning step involves defining which chemical to use in the jar test. Potential choices include the chemicals that are currently used in the treatment process, as well as alternative chemicals. The operator should have supplier data sheets for each chemical to be used in the jar test. Data should include, as a minimum, the following: chemical formula, specific gravity, percent weight, viscosity for liquids, solubility for solids, and safety information. If the chemical being evaluated in a test has never been used in the process, the previous information should be supplemented with data for unit cost, shipping and storage requirements, material safety data sheets (MSDSs), and reaction properties. Information about current or possible points of application is also necessary for successful simulation of the full-scale plant.

Current treatment performance data. In order to determine how well the jar testing conditions or procedures simulate conditions in the full-scale facility, the operator should obtain treatment plant performance data. This information can be related to that from the jar tests conducted under the same conditions.

Equipment

Jar test equipment consists of jars to hold the water, a mechanism to turn the impeller, the impeller, and lab equipment to analyze the results. The following sections describe each of these items and their role in the jar test procedure.

Types of Jars

The main function of the jar is to hold the water during the procedure so that observations can be made. Various types of containers have been used, including standard 1-L glass beakers and square 2-L containers. This section discusses the various types of containers and identifies advantages and disadvantages of each.

1-L circular. The 1-L glass beaker has been around the longest and is the smallest and least expensive container commonly used in jar tests. However, it has several disadvantages. First, it holds very little water. Consequently, a minor error in a chemical dosage can result in a large error in actual dosage, which will be further exaggerated when applied to a full-scale plant. Second, the stirring mechanism operating in the circular beaker causes the water to rotate with the paddles, reducing the amount of actual mixing during rapid mixing and flocculation. The 1-L beaker also provides a limited amount of water for evaluations. If several tests are to be run on the water after the jar test procedure, the sample may run out. Finally, the glass beaker does not provide for a good sampling point to develop settling velocity data. In general, the 1-L beakers should not be considered acceptable jars.

2-L circular. The 2-L glass beaker provides a larger volume of sample for the jar test and later analysis. The main disadvantage of the 2-L beaker is the circular shape, which also allows the water to rotate with the stirring mechanism. Stators can be placed in the 2-L beaker that will greatly reduce the rotation of the water, as shown in Figure 1-5. This arrangement allows the stirring mechanism to operate at a higher velocity gradient at a given speed than when stators are not used. A sampling siphon can also be added but with some effort. Care must be exercised when working with stators, however, because they often interfere with paddles and with sample withdrawal. These beakers are acceptable for jar testing if stators are added.

2-L square. A square 2-L beaker has been used recently as an alternative to the glass beakers previously discussed. Commercially available square jars are

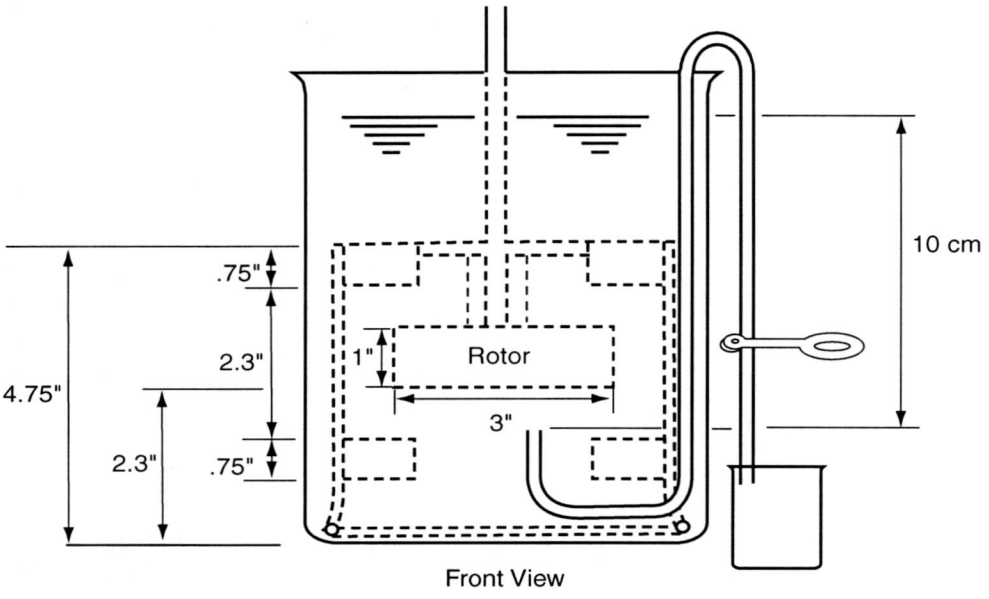

Front View

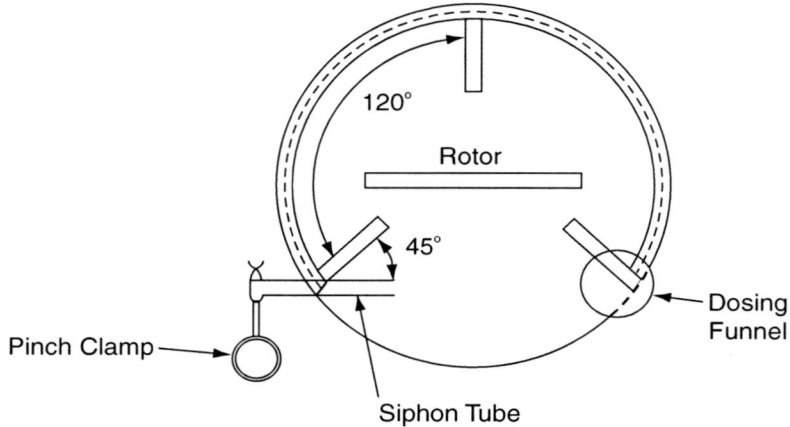

Source: Hudson H.E. Jr. 1981. Jar Testing and Utilization of Jar Test Data. In Water Clarification Processes: Practical Design and Evaluation. *Van Nostrand Reinhold Co., New York.*

Figure 1-5   Use of stators in circular 2-L beaker

fabricated by cementing together clear acrylic sheets. These jars are often referred to as *gator jars* (Cornwell and Bishop, 1983), because they were developed at the University of Florida. The square acrylic jars have a number of advantages over the glass beakers:

- The square configuration reduces the rotation of the water during mixing, eliminating the need for stators.

- The thicker wall and lower heat conductivity of the square acrylic beaker over the glass beaker reduce the change in water temperature during the jar test.

- Acrylic jars are less fragile and can be repaired if accidentally damaged.

- They can be equipped with a sampling port for collection of settling velocity data.

Stirrer

A stirring mechanism is needed to turn the impellers. Two basic types of stirring mechanisms are in use: gear-driven and magnetically driven units.

A gear-driven unit has a variable-speed motor that turns four to six gears. The motor(s) and gears are located above the jar test containers so that the shafts for the impellers extend down into each jar. Units are available that can be controlled from about 10 to 300 rpm.

The magnetically driven unit works on the same principle as a magnetic stirrer plate. The paddle contains a magnet that turns as the metal under it rotates.

Either type of unit may be used for jar testing. The magnetically driven unit has the advantage of providing open space above the jars to add chemicals. It is also possible to construct a jar stirrer using one or more variable-speed mixers. Units can be purchased that operate over a broad rpm range, allowing for an intensive rapid mix. They offer the advantage of testing different mix intensities simultaneously, because the stirrers are individually controlled. It may also be less expensive to buy three or four mixers and mount them on ring stands than to purchase commercially available units.

Impellers

Several types of impellers are available: paddle type, turbine, marine, and axial flow. Ideally, the impeller that best simulates the full-scale process should be used. Interchanging impellers requires a stirrer mechanism equipped to allow replacement of the shaft/impeller.

Magnetic stirrers are available only with flat-paddle impellers and would be difficult to retrofit with a different impeller type. They can, therefore, effectively simulate the action of a paddle, walking-beam, or flat-blade turbine type of mixing device. Figure 1-6 shows the velocity gradients for a standard magnetic stirrer in a 2-L square jar.

Standard top-mounting stirrers are generally provided with flat paddles. However, the shaft and impeller can be changed to allow testing of different impeller types. Figures 1-7 through 1-10 show G values for various combinations of jars and impellers.

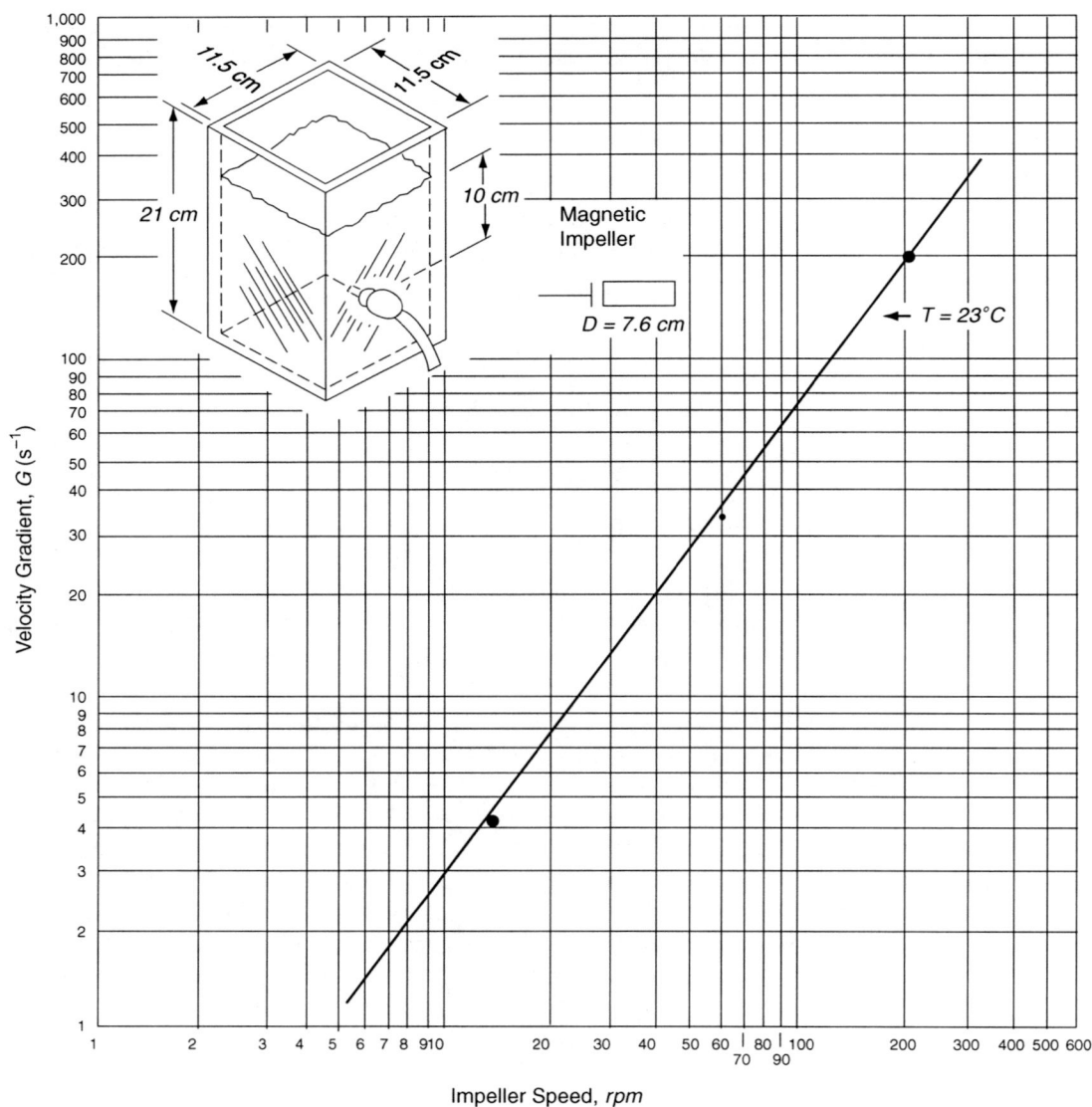

Source: Environmental Engineering & Technology, Inc. (personal communication).

Figure 1-6   Laboratory G curve for magnetic jar tester with gator jar

Water Bath

A water bath is a rectangular tank in which the jars sit during the jar test as raw water circulates around them to maintain the correct water temperature. The use of a water bath is not usually required unless the treated water is very cold. A simple test determines if a water bath is necessary: Take a sample of the raw water, and immediately run a jar test; take another sample of the same water and let it warm up 5 to 10 degrees, then conduct the same test and note if the results differ. Cold waters are frequently difficult to treat; therefore, even a slight temperature increase could make treatment easier than in the full plant, causing misleading results.

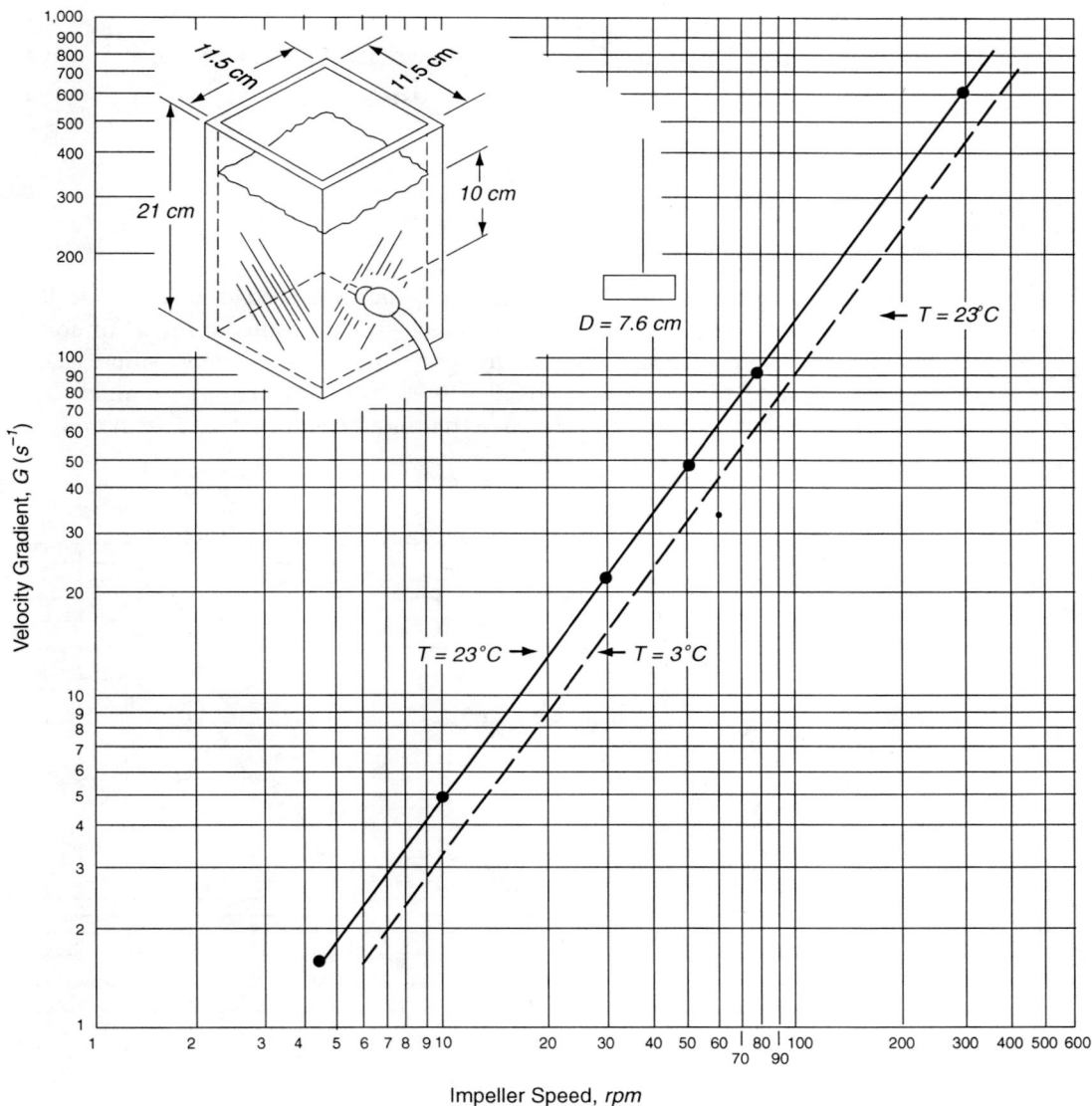

Source: Cornwell and Bishop (1983).

Figure 1-7   Laboratory G curve for flat paddle in the gator jar

Analytical Equipment and Laboratory Ware

A variety of lab equipment is required to prepare standardized solutions and analyze water samples. The most essential analytical equipment includes a turbidimeter, pH meter, thermometer, pipettes, burettes, and a laboratory weight scale. Compliance with new regulations may require operators to perform more advanced analytical analyses, which may call for the use of a carbon analyzer or UV-spectrophotometer. If the analytical equipment is not readily available, the operator must make arrangements with a commercial laboratory for needed tests. The laboratory may also provide suitable sample containers.

Chemicals

The chemicals used for pretreatment processes generally fall into one of four categories:

- Coagulants

- Coagulant aids

- pH controllers

- Oxidants/disinfectants

In addition to those listed, other chemicals may be needed during and after the jar test. Some of the chemicals listed are availabe as pure substances of uniform quality. Depending on their origins, others may vary in chemical composition due to their complexity, causing differences in effectiveness. Polymers are an example. In order to achieve useful results, jar testing must use treatment chemicals that are or will be used at the full-scale plant.

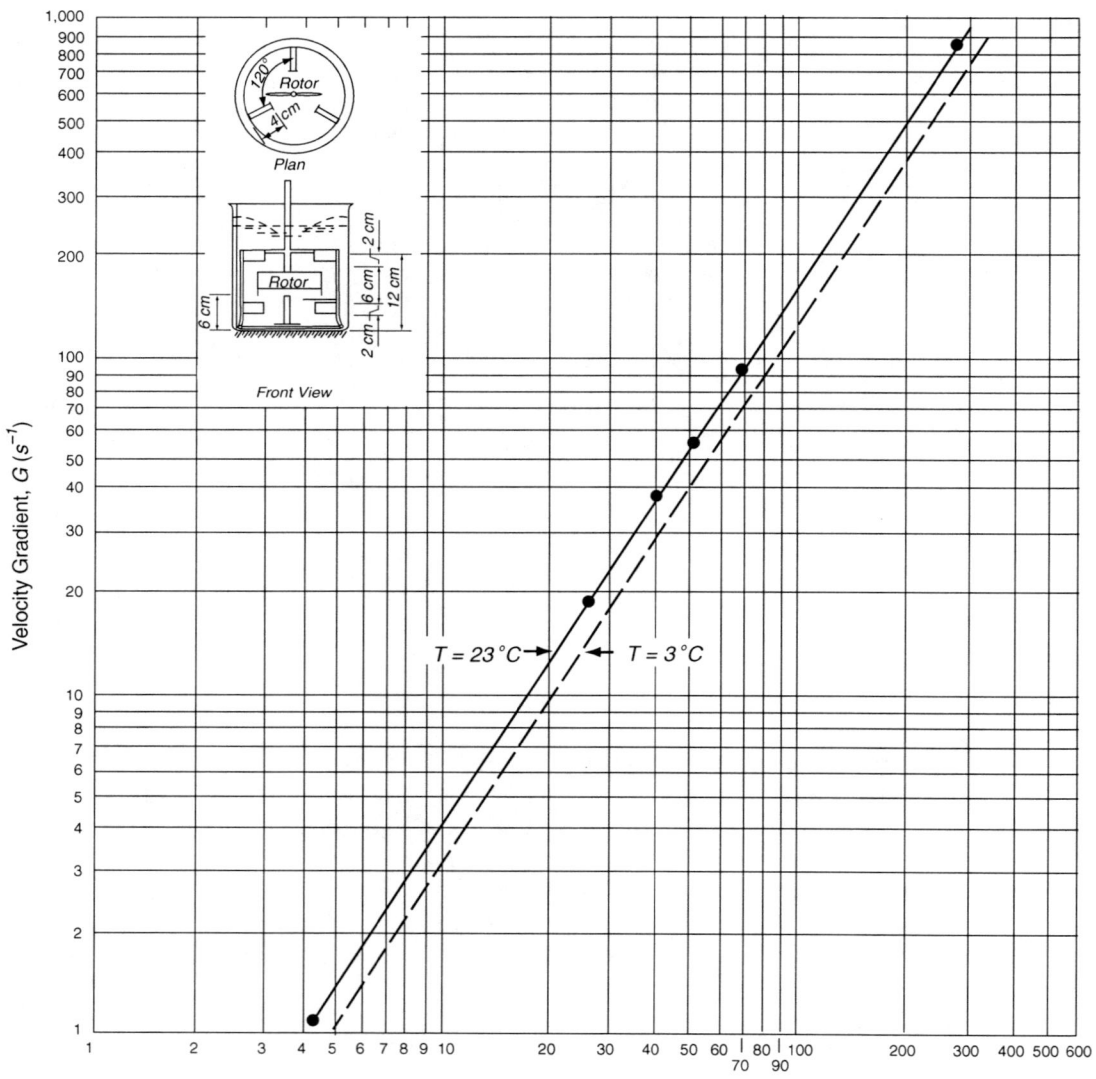

Source: Cornwell and Bishop (1983).

Figure 1-8   Laboratory G curve for flat paddle in the Hudson jar

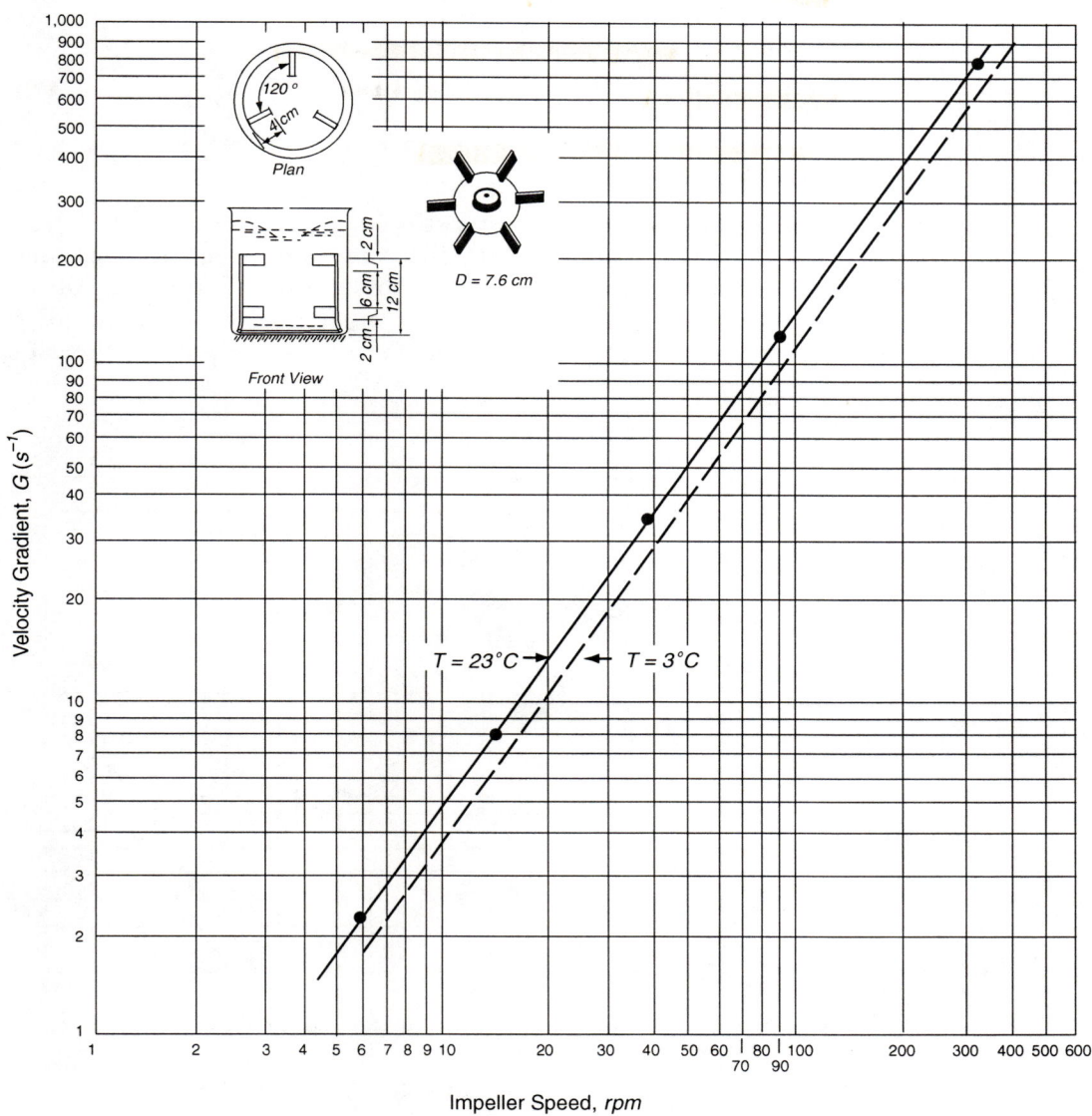

Source: Cornwell and Bishop (1983).

Figure 1-9 Laboratory G curve for flat-blade turbine in the Hudson jar

Coagulants

Loosely defined, a coagulant is a charged chemical species that is added to destabilize the substances contained in treated water so that they aggregate or adsorb to particles. The resulting floc can be removed by subsequent solids–liquid separation processes. Processes thought to cause particle and colloid destabilization include adsorption, charge neutralization, and enmeshment in a precipitate/coprecipitation. Positively charged hydrolysis products of metal salts may work by neutralizing the negative charges associated with naturally occurring particles. In other applications, when added at sufficient concentrations, these products form amorphous precipitates that enmesh or coprecipitate colloidal substances. The most commonly used coagulants include:

- **Aluminum sulfate (alum)**

- Ferric salts (ferric chloride and ferric sulfate)

- **Ferrous sulfate**

- Polymeric inorganic coagulants (partially neutralized metal salts such as polyaluminum chloride)

Expressing coagulant concentrations. When conducting a jar test, the correct doses of chemicals must be applied. The operator must understand whether the applied dose is stated on an "as product" or "as ingredient" basis. Much confusion can be avoided when doses/concentrations are expressed in terms of the molar amount of active ingredient, i.e., the mols of iron or aluminum per liter of solution.

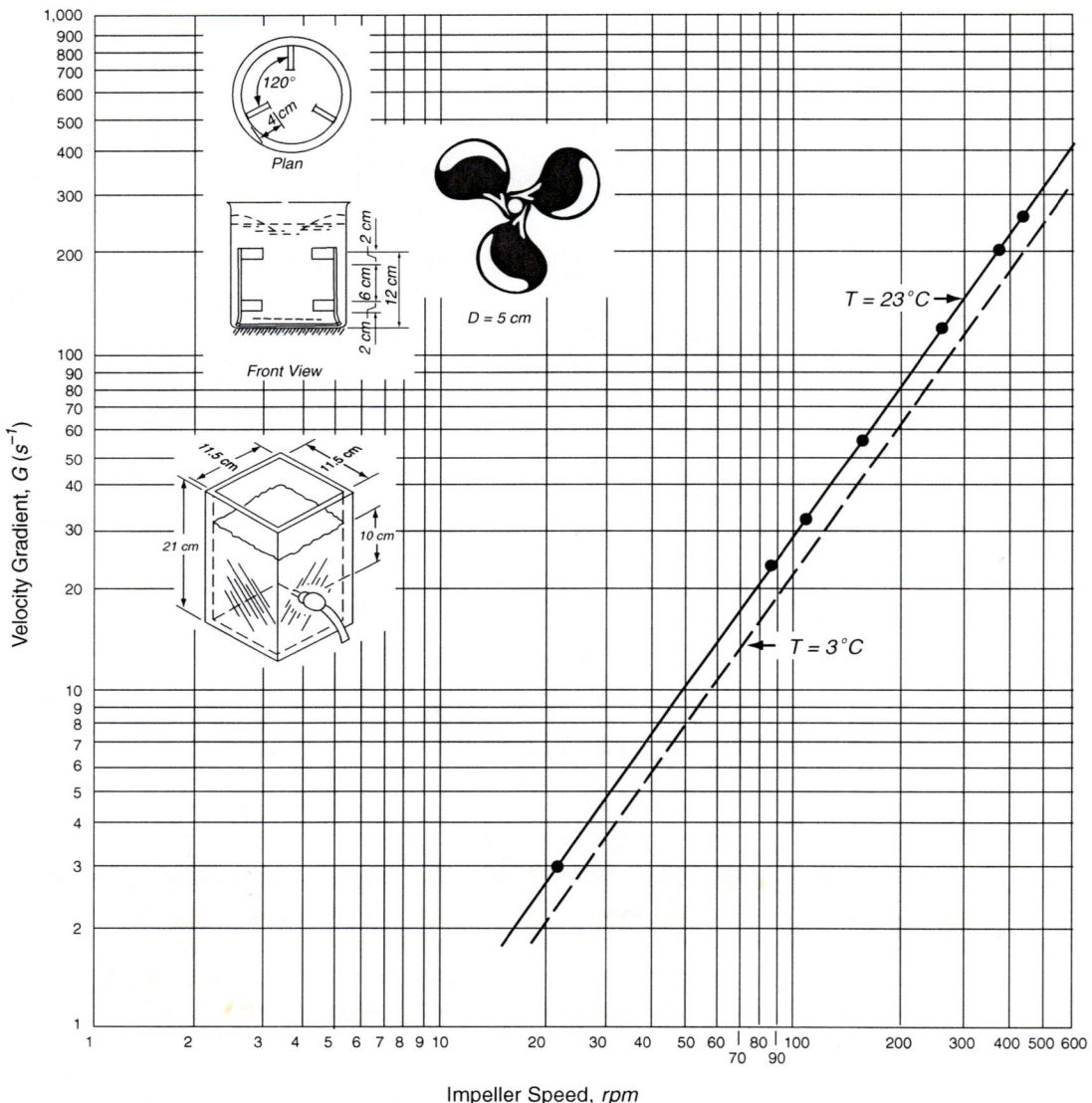

Source: Cornwell and Bishop (1983).

Figure 1-10   Laboratory *G* curve for marine propeller in either the Hudson or gator jar

This convention is of even greater importance for jar tests comparing different coagulants. Equimolar doses of aluminum and iron contain the same number of atoms. To determine the molar metal dose applied at the full-scale treatment plant, data for the plant flow rate, bulk chemical feed rate, and the chemical data sheet are needed. The datasheets for most liquid aluminum or iron-based coagulants report values for the specific gravity and %Al_2O_3 or %Fe. To convert to molar metal dose, apply the formula:

$$\text{Dose in micromol/L} = \frac{\text{Chemical fed rate (lb/day)}}{\text{Plant flow rate (mgd)} \times 8.34} \times \text{Factor} \qquad \text{(Eq 1-6)}$$

The factor depends on the coagulant used:

Coagulant	Factor
Dry alum ($Al_2(SO_4)_3 \cdot 14\ H_2O$)	3.365
Liquid aluminum product (alum, PACl, etc.)	% Al_2O_3 by weight/5.1
Dry ferric chloride (anhydrous, $FeCl_3$)	6.165
Dry ferric sulfate (anhydrous, $Fe_2(SO_4)_3$)	5.004
Dry ferrous sulfate (anhydrous, $FeSO_4$)	6.579
Liquid iron product (ferric sulfate, polyferric chloride, etc.)	% Fe by weight / 5.58

The molar metal dose at the full-scale plant obtained in this way can then be used as a baseline value for the jar test.

Example: A treatment plant feeds 5,000 lb of liquid alum per day at a flow rate of 5 mgd. According to the data sheet, the product contains 8.3 percent Al_2O_3 by weight. Therefore the dose equals:

$$\frac{5,000}{(5 \times 8.34)} \times \frac{8.3}{5.1} = 195\mu M \text{ (micromol/L) as Al} \qquad \text{(Eq 1-7)}$$

If the operator were to test ferric chloride as an alternative coagulant, a comparable dose to the currently applied alum dose would be 195 µM as Fe. Note that 195 µM as Al corresponds to 58 mg/L as alum; ferric chloride at a dose of 58 mg/L as Fe would be five times the molar Al concentration.

Making up stock solutions from the dry chemical. During jar testing, it is often inconvenient to feed a dry chemical to the jars. To test dry metal salts, stock solutions should be made up according to the following instructions:

Dry alum: Dissolve 10 g of dry aluminum sulfate ($Al_2(SO_4)_3 \cdot 14H_2O$) in distilled water and dilute to 1,000 mL. The resulting solution contains 10,000 mg/L alum, which corresponds to 0.17 percent Al_2O_3 by weight. Adding 1 mL of this stock solution to a 2-L jar results in a dose of 5 mg/L as alum or 17 µM as Al.

Dry ferric chloride (anhydrous): Dissolve 2.93 g of dry ferric chloride (anhydrous, $FeCl_3$) in distilled water and dilute to 1,000 mL. The resulting solution has a concentration of 1,000 mg/L or 0.1 percent Fe by weight. Therefore, adding 1 mL of stock solution to a 2-L jar results in 0.5 mg/L or 9 µM as Fe.

Dry ferric sulfate (anhydrous): Dissolve 3.57 g of dry ferric sulfate (anhydrous) in distilled water and dilute to 1,000 mL. The resulting solution has a concentration of 1,000 mg/L or 0.1 percent Fe by weight. Adding 1 mL of stock solution to a 2-L jar results in 0.5 mg/L or 9 µM as Fe.

Dry ferrous sulfate (anhydrous): Ferrous sulfate works in much the same way as ferric sulfate, except that it provides a bivalent iron ion (Fe^{2+}) when dissolved in water. Generally, the ferrous iron is oxidized to the trivalent form (Fe^{3+}) through chlorination prior to use. To oxidize ferrous sulfate to ferric iron, add 0.13 mg of chlorine for each milligram of ferrous sulfate. To prepare a stock solution, dissolve 2.7 g of ferrous sulfate (anhydrous) in distilled water and dilute to 1,000 mL. The resulting solution has a concentration of 1,000 mg/L or 0.1 percent Fe by weight. Adding 1 mL of stock solution to a 2-L jar results in 0.5 mg/L or 9 µM as Fe.

Note that the effectiveness of dry and liquid coagulants varies with dilution, and highly diluted stock solutions may degrade over time. Thus, new stock solutions must be prepared every day. Polymeric inorganic coagulants such as polyaluminum chloride can degrade very quickly and should not be diluted.

Making up stock solutions from liquid coagulants. Dilution of liquid coagulants is usually not required if the necessary volume can be added to test jars with a micropipette. If a micropipette is not available, dilution at a ratio of 1:100 by volume is recommended.

The amount in microliters (µL) to add to a 2-L solution can be calculated from the %Fe or %Al_2O_3 content and the specific gravity of the coagulant as follows:

For aluminum products:

$$\text{Volume (µL)} = \frac{\text{Dose, in µM} \times 5.10 \times \text{Volume, in L}}{\% \ Al_2O_3 \ \times \ \text{Specific gravity}} \times \text{Dilution factor} \qquad \text{(Eq 1-8)}$$

For iron products:

$$\text{Volume (µL)} = \frac{\text{Dose, in µM} \times 5.59 \times \text{Volume, in L}}{\%Fe \ \times \ \text{Specific gravity}} \times \text{Dilution factor} \qquad \text{(Eq 1-9)}$$

Figure 1-11 illustrates the relationship of molar metal concentration to alum or iron concentration in mg/L.

Example: A water utility tests two alternative coagulants, a polymeric ferric coagulant (sp. gr. 1.49, 12.27 percent Fe) and aluminum chlorohydrate (sp. gr. 1.34, 23.5 percent Al_2O_3). The plant currently uses anhydrous ferric chloride at a dose of 9.4 mg/L as Fe. Using Figure 1-11, the current dose corresponds to a molar metal concentration of 168 µM. To evaluate the effectiveness of these chemicals at equimolar concentrations, the following amounts have to be added to a 2-L jar:

$$\text{Aluminum chlorohydrate: } \frac{(168 \times 5.1 \times 2 \ )}{(23.5 \times 1.34)} \times 1 \ = \ 54 \ \text{µL} \qquad \text{(Eq 1-10)}$$

$$\text{Polyferric: } \frac{(168 \ \times 5.59 \ \times 2)}{(12.27 \times 1.49)} \times 1 \ = \ 103 \ \text{µL} \qquad \text{(Eq 1-11)}$$

If a micropipette is not available to dispense these small volumes, add 10 mL of the coagulant to a 1-L volumetric flask and dilute to the 1-L mark with distilled water (dilution factor = 100). The volumes to be added to a 2-L jar would then be:

$$\text{Aluminum chlorohydrate: } \frac{(168 \times 10.2)}{(23.5 \times 1.34)} \ \times 100 \ = \ 5,400 \ \text{µL} = 5.4 \ \text{mL} \qquad \text{(Eq 1-12)}$$

Polyferric: $\dfrac{(168 \times 11.17)}{(12.27 \times 1.49)} \times 100 = 10{,}300\ \mu L = 10.3\ mL$ (Eq 1-13)

To compare the ferric chloride side by side, a test would require 1 L of 0.5 percent Fe stock solution, made up by dissolving 14.5 g of the bulk chemical in distilled water. The specific gravity of the solution is assumed to be 1.0. The required volume for the jar test would be:

Ferric chloride: $\dfrac{(168 \times 11.17)}{(0.5 \times 1\ )} = 3{,}750\ \mu L\ 3.7\ mL$ (Eq 1-14)

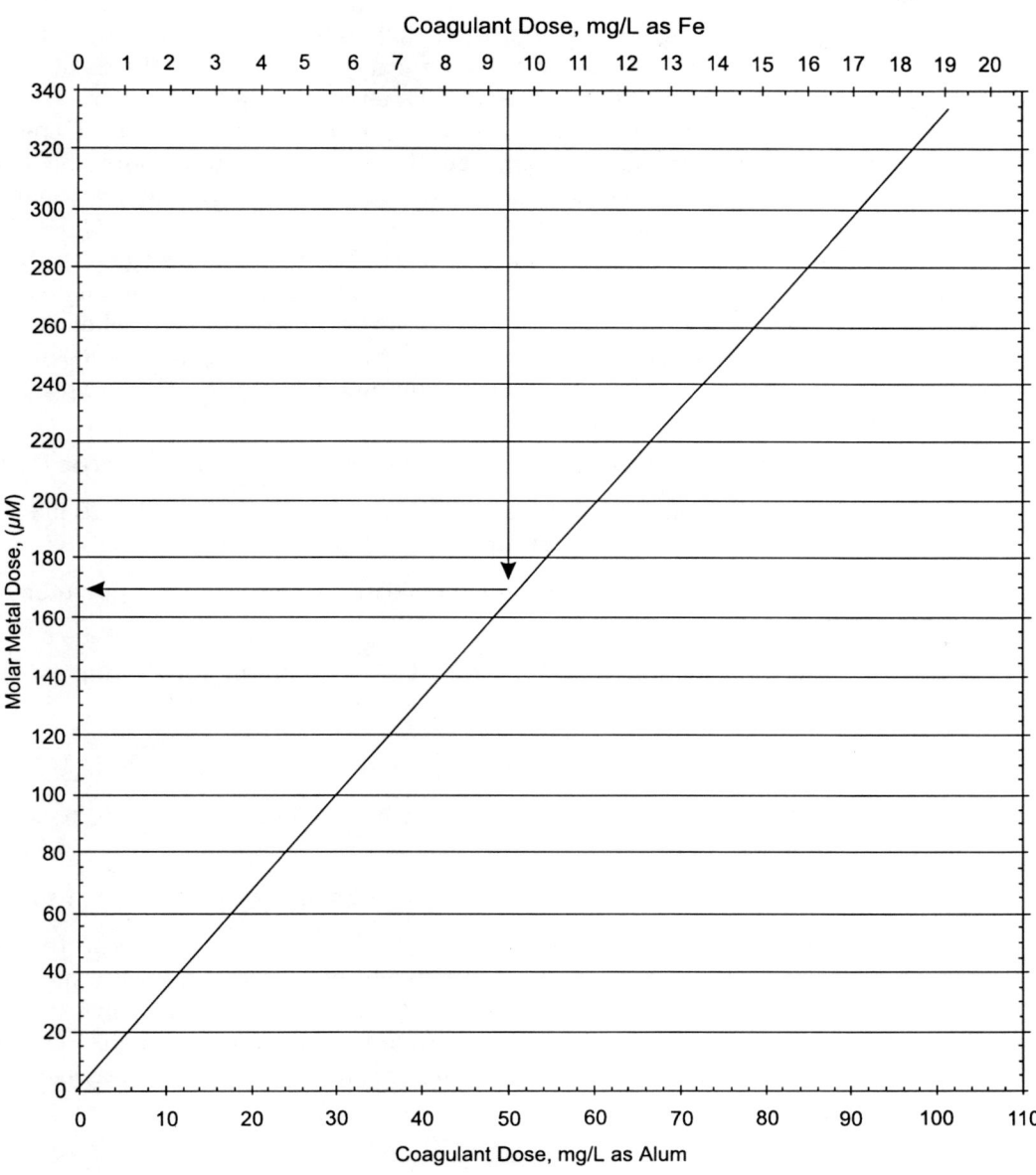

Figure 1-11   Expressing coagulant doses in molar metal concentrations

Table 1-1   Synthetic organic polymers

	Nonionic Polymer	Anionic Polymer	Cationic Polymer
Example	Polyacrylamide	Sodium polyacrylate	Polydiallyldimethyl ammonium chloride (polyDADMAC)
Characteristics	Neutral charge Used as floc and filter aid	Negative charge when dissolved Used primarily as floc aid	Positive charge when dissolved Used as either primary coagulant or coagulant aid

Coagulant Aids

In a most general sense, a coagulant aid is any substance used in conjunction with a primary coagulant, such as alum, to assist coagulation. By far the most significant coagulant aids are the synthetic organic polymers (Table 1-1). Polymers are chains of small subunits or monomers, and they may contain ionizable groups. Depending on the charges of the functional groups on the monomeric units and the molecular weights, they are classified as cationic, anionic, or nonionic polymers and of low, medium, or high molecular weight.

Guidance for the evaluation and selection of polymers is provided in the AWWA Research Foundation publication *Procedures Manual for Selection of Coagulant, Filtration, and Sludge Conditioning Aids in Water Treatment* (Dentel et al., 1986).

Preparation and use. Polymers are available as liquid, emulsion, and dry powder products. Instructions for preparation and use vary for these types.

Liquid and emulsion polymers:

1. Add 200 to 500 mL of distilled water to a clean 1-L volumetric flask.

2. After shaking the product container vigorously, weigh out 0.2 g of the polymer product onto an aluminum or plastic weighing dish.

3. Using a distilled water squeeze bottle, rinse all of the polymer into the volumetric flask.

4. Fill the volumetric flask to the 1-L mark with distilled water.

5. Cap and shake for at least 1 min.

6. The strength of the stock solution will be 200 mg/L or 0.2 mg/mL.

7. Therefore, 1 mL of the stock solution added to a 2-L jar will be equivalent to a dose of 0.1 mg/L.

8. Most polymer stock solutions of this strength will degrade within 24 h.

In order to compare the performance of a liquid polymer product to that of a dry product, the polymer content of the liquid should be obtained from the manufacturer. Many liquid polymer products contain only 20 to 40 percent of actual polymer. Therefore, for a 20 percent solution, a dose of 1 mg/mL of liquid polymer product contains only 0.2 mg/L of actual polymer. Dry polymers are essentially 100 percent polymer.

Dry polymers:

1. Add 200 to 500 mL of distilled water to a clean volumetric flask.

2. Drop in a magnetic stir bar, and place the flask on a magnetic stirrer.

3. Weigh out 0.2 g of the polymer onto an aluminum or plastic weighing dish.

4. Using a distilled water squeeze bottle, rinse all of the polymer into the volumetric flask. Mix at medium speed for at least 2 h.

5. Remove the stir bar, and fill the volumetric flask to the 1-L mark with distilled water.

6. Cap and shake it for at least 1 min.

7. The strength of the stock solution will be 200 mg/L or 0.2 mg/mL.

8. Therefore, 1 mL of the stock solution added to a 2-L jar will be equivalent to a dose of 0.1 mg/L.

9. Most polymer stock solutions of this strength will degrade within 24 h.

Polymers are expensive, but they usually work well at low concentrations. The maximum dose should be less than 1.5 mg/L, and doses as small as 0.05 mg/L may prove effective as aids for coagulation. To ensure compliance with dose limits set by the US Environmental Protection Agency, consult suppliers to find a polymer for a range of recommended doses. Otherwise, a good range of doses to test first may be 0, 0.1, 0.25, 0.5, 0.75, and 1.0 mg/L. The proper volume of polymer solution can then be calculated for each jar. For example, to test a polymer solution at a concentration of 0.1 g/L (0.1 mg/mL), the amounts of polymer to be added for the doses mentioned would be 0, 2, 5, 10, 15, and 20 mL for a 2-L jar. Once these doses are evaluated, further jar tests can be performed with different doses to more exactly identify the optimal polymer dose.

pH Control

Control of pH during jar testing is important for a number of reasons:

- Simulating existing conditions
- Evaluating alternatives for optimum turbidity removal
- Evaluating alternatives for optimum DOC/THM precursor removal
- Minimizing coagulant residual in the distribution system

If the coagulation reaction using aluminum or iron salts as the primary coagulants occurs at nonoptimized pH conditions, the quality of the treated and filtered water will be degraded by undesirable concentrations of dissolved aluminum and iron. The metallic salt coagulants are generally more susceptible than the polymers to loss of effectiveness under nonoptimum pH conditions.

Often, pH is controlled only by the dosage of the coagulant applied. If the water treatment plant operates this way, jar tests should also be conducted the same way to find the optimum chemical dosage for routine treatment of water.

Sometimes, however, the properties of a raw water cause optimum treatment to occur at a pH significantly different from that obtainable from the coagulant alone. A jar test series will quickly demonstrate this condition. In some cases, cost-effective treatment would adjust the raw water pH with another chemical (acid or base) to achieve the best pH for the coagulant of choice. Again, the jar test is perhaps the most valuable tool for rapidly determining the best combination of coagulants and other chemicals to achieve the most cost-effective and process-efficient reaction pH.

Adjusting pH during a jar test can be a hectic task involving measurements of and adjustments to pH in six jars during a 1-min rapid mix. The easiest and generally most accurate method is to predetermine the required acid or base dose by taking a small sample of raw water, say 100 to 200 mL, and adding the coagulant dose equivalent to the dose that will be added to the 2-L jar. Place the sample on a standard stirrer and titrate with acid or base, recording the dose to reach the desired pH level. A single titration can determine the acid or base doses to reach different pH levels for one coagulant dose. To determine the acid or base doses for different coagulant concentrations, one must repeat the test. With this information, the proper volume can be premeasured and fed to the jar during, before, or after coagulant addition. The pH at the end of the rapid mix phase still should be measured and recorded.

Common chemicals.   A number of chemicals are routinely used for pH control:

- Lime
- Sodium hydroxide (caustic)
- Hydrochloric acid
- Sulfuric acid
- Carbon dioxide
- Soda ash

Historically, because water treatment has focused on turbidity removal, lime and caustic have been the most widely used. Currently, however, increasingly stringent disinfection requirements in the Surface Water Treatment Rule (SWTR) and regulations for disinfection by-products (DBPs) are causing a shift toward maximizing DOC/DBP precursor removal. Because optimum organics removal tends to occur at relatively low pH values, the use of acid to depress pH is increasing.

In many cases, the use of caustic is more convenient for jar testing. Lime is dosed in a suspension that requires continuous stirring. Caustic can be used as a dissolved liquid without stirring. The +2 charged calcium ions associated with lime have been shown to act as a coagulant aid. Consequently, its use for increasing pH may have a slightly more beneficial effect than the use of caustic, as the sodium ion has only a single positive charge. However, in many cases, the difference provides negligible benefits. If a base is required for pH adjustment in a jar test, either is generally acceptable, regardless of what the full-scale facility is using, and caustic is much easier to use.

Once the dose of lime or caustic has been established, the following conversion factors can be used to adjust values if the water treatment plant and the jar test series use different products:

- mg/L CaO = mg/L $CaCO_3$ × 0.56
- mg/L $Ca(OH)_2$ = mg/L $CaCO_3$ × 0.74
- mg/L Na_2CO_3 = mg/L $CaCO_3$ × 1.06
- mg/L $NaOH$ = mg/L $CaCO_3$ × 0.80

Sodium hydroxide (caustic soda): For jar testing, a 0.1N solution of sodium hydroxide is generally sufficient for pH adjustment. Reagent-grade sodium hydroxide is usually supplied in pellet form. To prepare a 0.1N solution:

1. Add 200 to 500 mL of distilled water to a clean 1-L volumetric flask.

2. Drop in magnetic stir bar, and place the flask on a magnetic stirrer.

3. Weigh out 4 g of the sodium hydroxide pellets onto an aluminum or plastic weighing dish.

4. Pour all of the pellets into the volumetric flask. Mix at medium speed until all of the pellets dissolve.

5. Remove the magnetic stir bar and fill the volumetric flask to the 1-L mark with distilled water.

6. Cap and shake it for at least 1 min.

The strength of the stock solution will be 4,000 mg/L, or 4 mg/mL as 100 percent NaOH. Therefore, 1 mL added to a 2-L jar will be equivalent to a dose of 2 mg/L.

Oxidants and Disinfectants

In the last several years, interest has increased in the use of alternative oxidants and disinfectants. The reasons for this increasing interest include the need to control DBPs and the need to provide adequate disinfection to satisfy new regulations. Bench-scale testing can provide a way to screen the effectiveness of alternative oxidants and their impact on disinfectant by-product formation.

Chlorine, chlorine dioxide, ozone, and potassium permanganate are the chemicals most commonly used as primary disinfectants and oxidants. Chloramines are considered a weak oxidant and generally not suitable for primary disinfection. Rather, chloramines are most often used as a secondary disinfectant to provide residual disinfection in the distribution system. Advanced oxidation processes include the use of hydrogen peroxide and UV radiation in combination with ozone.

Chlorine. Chlorine is the most common oxidant/disinfectant in the water industry. It is usually available in liquid or gaseous form in pressurized metal tanks, as a concentrated aqueous solution (sodium hypochlorite), or as a solid (calcium hypochlorite). Whatever form is added, chlorine disproportionates into Cl_2, HOCl (hypochlorous acid), and OCl^- (hypochlorite ion). Addition of liquefied or gaseous chlorine decreases pH and alkalinity, while applying hypochlorite increases them. Because certain risks accompany the use of chlorine gas, sodium hypochlorite solution is recommended for jar testing. Sodium hypochlorite is widely available as common household bleach. If this concern arises, laboratory grade NaOCl should be used. Of course, the same chemical may also be available from the full-scale plant.

The chlorine content of commercial sodium hypochlorite solutions is often expressed in percent by weight (%w/v). Assuming, that the specific gravity of the liquid is 1, a percent by weight value can easily be converted to a mass concentration as Cl_2 by multiplying by 10,000. Depending on the concentration of the chemical, an appropriately diluted stock solution may have to be prepared. For example, a laboratory grade NaOCl-solution contains 5 percent Cl_2 by weight. This figure corresponds to 50,000 mg/L as Cl_2. Add 100 mL of the solution to a 1-L volumetric flask and fill to the 1-L mark using distilled water. The resulting stock solution contains 5,000 mg/L or 5 mg/mL chlorine, therefore 2 mL added to a 2-L test jar equals a 5 mg/L dose.

It is recommended to verify the chlorine content by a standard analytical method.

Chlorine dioxide. This disinfectant is generated continuously in treatment plants, because it rapidly decomposes. Problems associated with bench-scale evaluations focus mainly on obtaining a high-quality stock solution. If the chemical cannot be obtained from the treatment plant for immediate use, it is recommended to generate it in the laboratory as outlined in *Standard Methods* (APHA, AWWA, and WEF, 1998). When applying the chemical during jar testing, a gas-tight syringe is recommended because of the tendency of the chlorine dioxide to degas from the solution. In addition, the stock solution should be standardized prior to application.

Ozone. This disinfectant also must be generated on site because of its instability. A typical ozonation system consists of a generating unit, where the ozone is produced by electrolytic oxidation, and a contactor. The use of ozone is limited in jar testing, for it cannot be applied to the jar as a concentrated solution. Therefore, the raw water must be ozonated prior to transfer to the jars using a continuous flow reactor. Ozonation of settled water is thus not usually considered practical in jar testing.

Potassium permanganate ($KMnO_4$). This moderately strong oxidant does not cause THM-formation, but it may result in pink water at high doses. For jar testing, a permanganate solution can be prepared from the powered or granular solid. Dissolve 1 g of potassium permanganate ($KMnO_4$) in distilled water and dilute to 1,000 mL. The resulting solution contains 1,000 mg/L or 1 mg/mL potassium permanganate, so adding 1 mL of the stock solution to 2 L of raw water will result in a 0.5 mg/L dose. When evaluating potassium permanganate as an alternative oxidant, it is recommended to determine the appropriate dose before beginning jar tests by adding various doses to beakers containing the raw water and selecting the dose that does not result in pink water after a preset period of time, e.g., 30 min.

Chloramines. This weak oxidant forms a persistent residual, making it a suitable secondary disinfectant. Chloramines are formed by the reaction of ammonia with aqueous chlorine. Jar tests may be used to evaluate how the point of ammonia application affects disinfectant by-product formation. Although ammonia can be applied in various forms in full-scale treatment, for jar testing it is convenient to make up a solution from ammonium sulfate or chloride. For a 1,000 mg/L NH_3-N solution, dissolve 3.82 g of anhydrous ammonium chloride or 4.72 g of anhydrous ammonium sulfate in distilled water and dilute to 1 L.

Advanced oxidation processes (AOP). Treatment methods such as ozone/UV and ozone/hydrogen peroxide processes show promise as alternatives to traditional oxidants, but they are relatively expensive and not widely used. AOP are commonly evaluated in pilot-scale tests.

Other Chemicals

Successful simulation of full-scale plant conditions requires addition of any other chemicals normally added during rapid mix, flocculation, or sedimentation at the same doses in the jar tests. These additional chemicals may have profound effects on treatment performance. The list of products may include hydrofluosilicic acid (a dental caries prophylactic) or hexametaphosphate (a corrosion inhibitor).

If testing includes laboratory analyses that cannot be conducted immediately, samples must be preserved. Preparations for a jar test must ensure that the appropriate preservatives are on hand. Commonly used preservatives may include nitric acid for metal analyses, sodium thiosulfate for quenching DBP samples, or ethylenediamine for preserving by-products of chlorine dioxide disinfection. Refer to documentation for the appropriate method or consult a laboratory analyst.

Documentation

Practices for collecting jar test results should not be overlooked. During jar testing, several events may occur simultaneously, and the actual evaluation of the test results may not be made until several days after the test are performed. Significant observations may be difficult to recall if not written down. This documentation requires preparation of data sheets and test protocols, which serve not only for data collection but also as reminders for the steps to take during a jar test. The data sheets should therefore be designed so that the form holds all relevant information for the jar test in the sequence that the data is collected.

Date:		Time:		Source Water					
		Concentration (mg/L)		pH	Turbidity (ntu)	Alkalinity (mg/L as CaCO$_3$)	UV254 (cm^{-1})	TOC (mg/L)	
Coagulant:									
Oxidant									
Polymer:									

Jar Number			1	2	3	4	5	6
Rapid Mix		G (s^{-1})						
		rpm						
		Duration (s)						
Flocculation		G (s^{-1})						
		rpm						
		Duration (s)						
Coagulant Dose (mg/L)								
Volume of Coagulant Added (mg/L)								
Oxidant Dose (mg/L)								
Volume of Oxidant Added (mL)								
Polymer Dose								
Volume of Polymer Added (mL)								
Coagulation pH								
Settling Velocity (cm/min)	Depth of Sampling (cm)	Time of Settling (min)	Turbidity (ntu)	Turbidity (ntu)	Turbidity (ntu)	Turbidity (ntu)	Turbidity (ntu)	Turbidity (ntu)
TOC (mg/L)								
UV-254 (cm^{-1})								

Figure 1-12   Sample data sheet

Figure 1-12 shows an example data sheet. In preparing for the jar test, it allows entry of source water quality parameters, concentrations of the chemicals, and the hydraulic characteristics of the full-scale plant and corresponding jar test parameters.

It is preferred to prepare these data sheets using computer spreadsheet programs. These tools allow both rapid calculation of removal efficiencies and capabilities for graphical representation of results.

CONDUCTING THE JAR TEST

If all the preparation tasks described in the previous section have been performed, the actual jar test can be conducted. At this point, the operator should have defined the study goals (e.g., optimizing coagulant or polymer dose) and the testing parameters (hydraulic characteristics of the plant, points of application, etc.). Testing and analytical equipment should be ready (jars, stirrer, properly labeled sample containers, turbidity meter, pipettes, etc.). All reagent solutions should be prepared (coagulants, polymers, oxidants, etc.), and a data sheet should be available in a convenient spot.

The operator should ensure that all chemicals are properly labeled and that the reagent solutions are thoroughly mixed. The reagent containers should be placed near the jar test equipment in the order that they are used. Glass pipettes should be labeled, too, and placed in or next to the corresponding reagent containers. Automatic pipetters should be set to the correct volumes.

The jars and the paddles of the stirring mechanism should be cleaned by wiping with a damp cloth and rinsing with warm tap water to remove any residue from previous jar tests.

The data sheet should be a good guide for conducting the jar test:

1. Treatment performance data is often expressed in terms of percentage removal. Therefore an important beginning step is to determine the quality of the raw water to be tested. Such data may be obtained from treatment plant records or determined during the jar test. Also, pH and alkalinity data may help determine necessary additions of acid or base.

2. Enter the names and concentrations of the chemicals to be added on the data sheet. This information is necessary to determine the volumes of chemical to be added during the jar test.

3. Enter the G values for the rapid mix and flocculation stages of the full-scale plant on the data sheet. If the effect of varying mixing intensities is to be evaluated, use the appropriate range of G values. Convert these values to the appropriate rpm in the jar test. Depending on the jar test equipment used, refer to Figures 1-6 through 1-10 to determine the correct rpm value.

4. Enter the detention times for the rapid mix and the flocculation stages of the full-scale plant in the data sheet. If the effects of detention times are to be evaluated, use an appropriate range of durations.

5. Enter the coagulant doses on the data sheet. If the test will determine an optimum coagulant dose, it is useful to select doses in increments of 10 mg/L for alum, or equivalent doses if other coagulants are to be tested. Smaller increments may be used for fine-tuning the optimum coagulant dose. Then calculate the volume of coagulant to be dispensed into the jar to obtain the desired coagulant dose.

6. If applicable, proceed with all the other chemicals in a similar manner.

7. Based on the surface loading or overflow rate of the sedimentation basin of the full-scale plant, determine the critical settling velocity as outlined in the discussion of sedimentation in the General Theory section. Divide the depth of the sampling port on the jars in centimeters by the settling velocity to obtain the sampling time in minutes.

8. Fill the jars with the water to be tested, and position them under the stirring apparatus so they are centered with respect to the impeller shafts.

9. Lower the impellers or paddles so that they are about one-third from the bottoms of the jars.

10. Begin the flash mix period based on the previously determined values. Do not forget to record the starting time. Dispense the desired doses of chemicals as rapidly as possible into the jars. Dispense the chemicals in the same sequence as at the full-scale plant, unless the effect of moving the point of application is to be evaluated.

11. After the rapid mix period, decrease the mixing speed to the predetermined value for the flocculation period. At this point, the coagulation pH is typically measured.

12. After the flocculation period, stop the mixer and remove the paddles from the jars. Collect samples at the times previously calculated to simulate the

full-scale sedimentation basin. Sample withdrawal may be accomplished either by the use of a syringe, a fixed sampling port, or a pipette. The first portion of sample taken from a fixed port should be discarded. When using a syringe, samples should be taken from the same depth as the fixed port.

13. After sampling, conduct the laboratory analysis, observing holding times required for specific analytical applications.

14. Enter laboratory results on the data sheet.

INTERPRETING THE RESULTS

Successful interpretation of jar test results is possible only when adequate data have been collected and recorded. Comparisons of jar test data require such key information as all water quality data, hydraulic data, and the types and doses of the chemicals used. Once all the data are collected, the easiest way to evaluate them is to prepare charts and graphs using computer spreadsheet programs. Well-prepared charts and graphs show how well the jar tests simulate the full-scale plant as well as the effects of alternative treatment options.

Guidelines for Charts

When preparing graphs and charts, the operator should consider a number of guidelines.

Percentage removal versus absolute values. Jar test data can be easily converted to percentage removal values using the formula:

$$\text{Percentage removal} = (1 - \text{Final value/Initial value}) \times 100 \qquad \text{(Eq 1-15)}$$

This method may be chosen to assess treatment performance when initial water quality parameters vary or no absolute concentration is specified as the treatment goal.

Use of percentage removal data may obscure actual differences, so the initial values or ranges of values should be included in the data report. For example, consider the following data:

	Jar Test 1	Jar Test 2
Source water turbidity (ntu)	18.0	7.0
Settled water turbidity (ntu)	2.2	1.0
Percentage removal	87.8	85.7

Although the percentage removal is higher in jar test 1, the hypothetical treatment goal of 1.0 ntu in the settled water is reached only in jar test 2.

Logarithmic scales. When the data to be displayed in a set of charts include values in increasing increments, such as 0.05, 0.2, 0.5, 5, 20, 100, a logarithmic scale may prove helpful to compress the distances between the data points.

'Smoothing" functions. Modern spreadsheet programs can display data by connecting the data points with a smooth curve. However, the operator must keep in mind that these curves do not represent the actual data, so they may lead to misinterpretation of the results. It is recommended that the operator plot the actual data points and use straight lines to connect the points to improve readability.

Trendlines.   When using a computer to graph jar test data, the program may allow display of a trendline or "best-fit" line. This option calls for a careful choice of the correct statistical method for calculating the trendline to avoid a misleading result. In many cases a trendline can be well approximated by just "eyeballing" the data points. The use of trendlines is especially helpful when many data points are displayed in the same plot.

Error bars.   If the accuracy of the analytical method is known, instead of just the data point, error bars can be included in the graph. This addition allows an assessment of whether two data points are significantly different from one another.

Types of Charts

Several methods are available for reporting or reviewing the results of jar testing. The graph type selected depends on the preference of the operator and the intended use.

A bar graph is suitable only for comparing a few values in different categories in a single plot. Figure 1-13 represents a comparison of jar testing results with the results from a full-scale plant. Note that the percentage removal values for turbidity, UV absorbance, DOC, and THMFP were comparable between the full-scale plant and the jar tests. In addition, similar amounts of THMs were formed as a result of prechlorination. These results indicate that the jar testing protocol successfully simulated the full-scale treatment plant. The bar graph in Figure 1-14 demonstrates how jar test results may aid in selecting a suitable synthetic organic polymer for enhanced solids–liquid separation. The chart indicates that nonionic polyacrylamide performed best in the jar tests at all three sampling times, i.e., overflow rates.

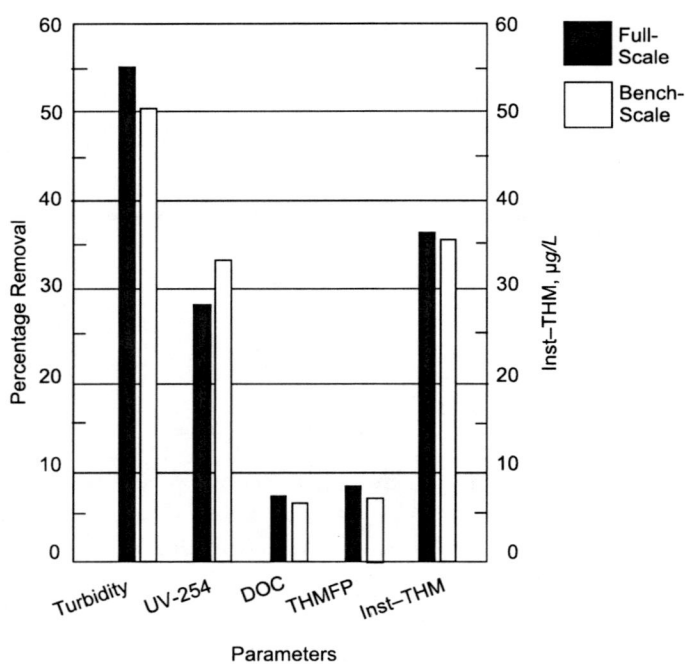

Source: City of Phoenix, Arizona (1989).

Figure 1-13   Example correlation between jar test results and full-scale performance

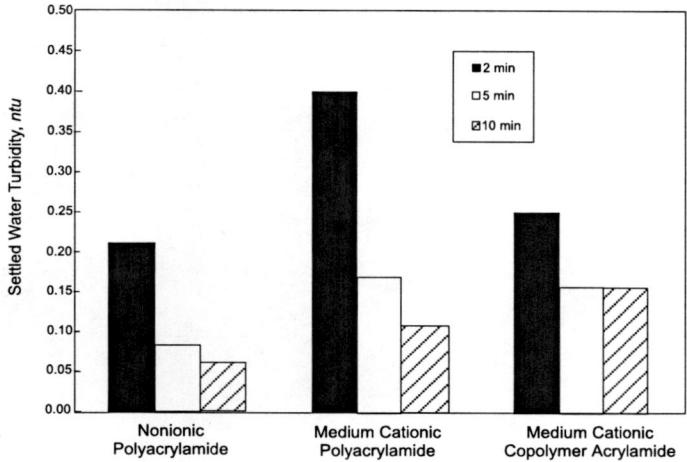

Source: AH Environmental Consultants, Inc. (1997).

Figure 1-14   Use of the jar test to select coagulant aids: Turbidity versus settling time

An *x-y* (scatterplot) graph is suitable for displaying a series of data points. Multiple lines can be used to express results under various conditions. For example, in Figure 1-15, UV-absorbance data were collected from six jars that received increasing alum doses. The test was then repeated at a different pH. This graph shows UV-254 absorbance versus dose, and each line represents the data series for one pH level. The plot indicates that coagulation at pH 5.8 yielded lower residual UV-254 absorbance than coagulation at pH 6.6. Another example of an *x-y* graph, shown in Figure 1-16, illustrates the effect of flocculation time on settled water turbidity at two different settling velocities, i.e., overflow rates. The graph in Figure 1-17 was used to determine the optimum polymer dose for a treatment plant based on jar tests. The logarithmic scale on the *y*-axis helps illustrate the differences in turbidity.

Topographs can be used to simulate a three-dimensional view of the collected data. They are commonly used to represent large data sets, and they often require special software packages. Figure 1-18 shows a topograph of THM formation potential as a function of applied ozone and alum dose.

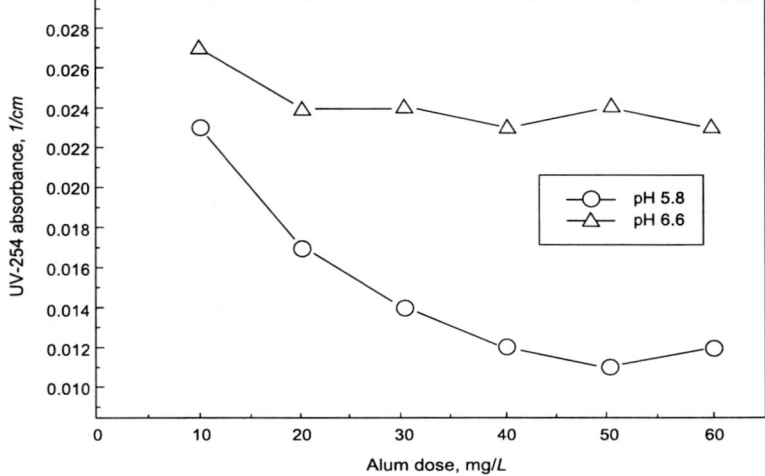

Source: AH Environmental Consultants, Inc. (1997).

Figure 1-15   Use of the jar test to optimize the coagulation pH: UV-254 versus alum dose

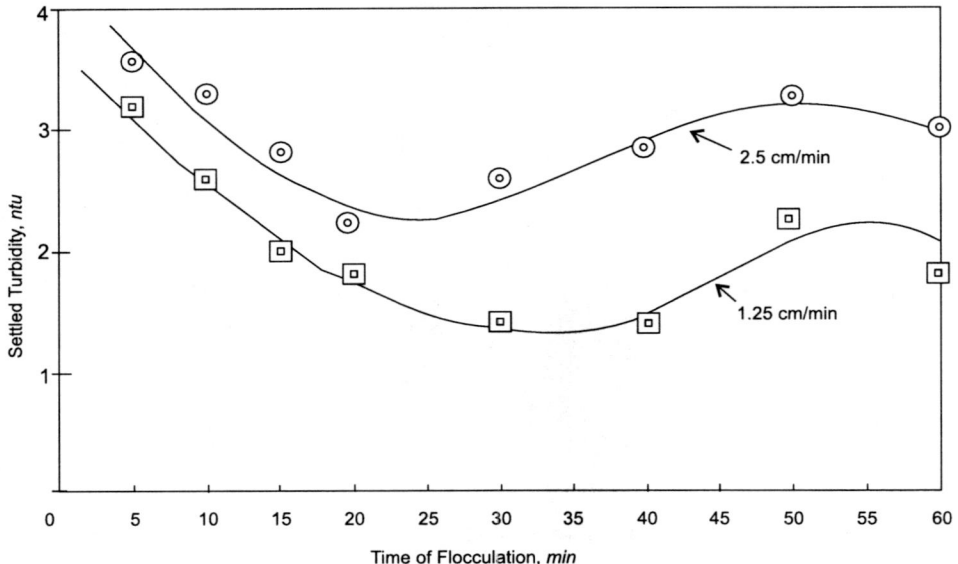

Source: Environmental Science & Engineering, Inc. (1982).

Figure 1-16   Example use of the jar test: Flocculation time versus settled turbidity

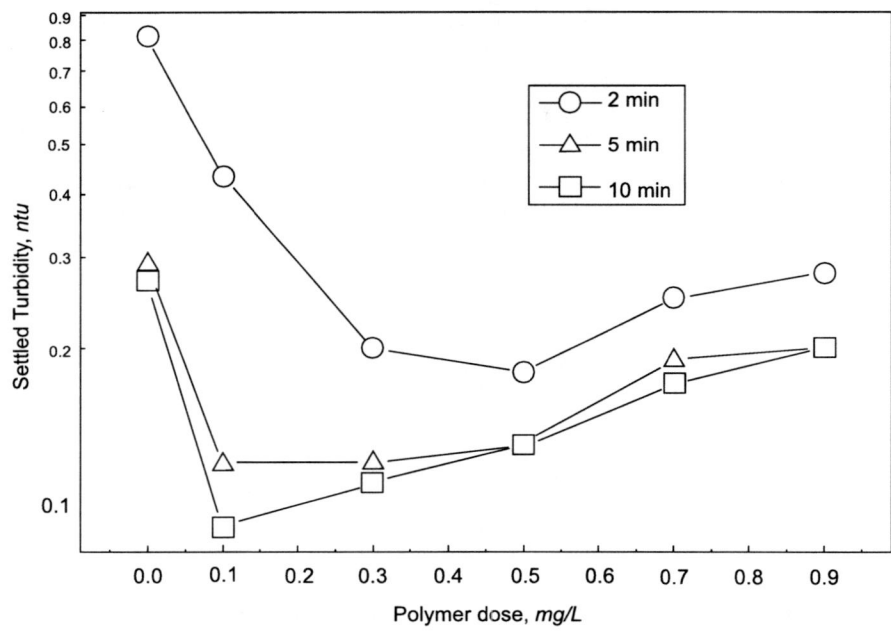

Source: AH Environmental Consultants, Inc. (1997).

Figure 1-17   Use of the jar test to determine the optimum polymer dose: Turbidity versus dose and settling time

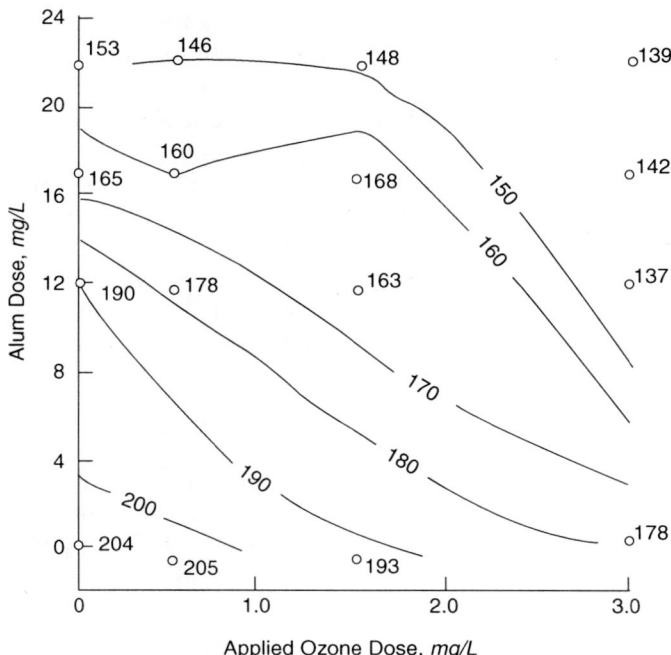

Source: Environmental Engineering & Technology, Inc. (1987).

Figure 1-18   Alum dose, ozone dose, THMFP topograph

SPECIAL APPLICATIONS

Bench-Scale Evaluation of Filtration

The purpose of a filter is to remove particles from the water that flows through it. Current theories indicate that effective particle removal in filtration depends on both physical and chemical factors. Effective filter performance depends not only on the standard design criteria of filtration rate, media size, and bed depth, but also on

- Transport of the particles to the surfaces of the filter media

- Attachment of the particles to the media surfaces

- Concentration and size of the particles applied to the filters

Therefore, filtration is a difficult water treatment process to evaluate using bench-scale techniques. Consequently, the similarity between laboratory and full-scale filter performance cannot always be assured. However, bench-scale filtration techniques are sufficient to determine some of the required filtered water quality information and to establish trends in performance.

Available bench-scale filtration alternatives include

- 0.45 μm membrane filters

- 1 μm glass fiber filters

- Whatman 40 filters (8 μm)

Choice of bench-scale filter test apparatus. The selection of a specific method depends on the information required and the ability of the method to produce filtered water quality similar to that from the full-scale filters. For example, based on standards for 0.45 μm membrane filters, any natural organic material that passes through the filter can be operationally defined as in the dissolved phase rather than the particulate phase. Consequently, if the objective of the study/testing program is to determine the efficiency of the treatment process at converting the dissolved organic material to the particulate phase, the 0.45 μm filters would be recommended.

For some studies, the objective is to evaluate the impacts of various treatment alternatives on filtered water quality. Typical water quality parameters of concern include color, chlorine demand, TOC, and THMFP. The 1 μm glass fiber filter is usually sufficient to meet these objectives.

Whatman 40 (8 μm) filters are most useful for comparing filtering effectiveness, such as comparing the use of filter aids or conducting bench-scale direct-filtration studies. However, head loss data collected in this way may be misleading.

Water From Rapid Mix

This procedure is applicable for evaluating water after coagulants such as alum or ferric chloride have been added but before polymer addition. This test is useful only to evaluate efforts to improve water quality using a polymer without modifying the alum or ferric dose. The procedure is similar to the one given previously, but is easier to perform, as the water to be tested has already been dosed with one or more chemicals during rapid mix. All steps are the same, except that coagulant addition is omitted. If chemicals for pH control have been added in the plant prior to the sampling point, these steps are also omitted.

Decreasing the Primary Coagulant With Coagulant Aids

A test decreasing the primary coagulant with coagulant aids is applicable to approximate the reduction in primary coagulant as a result of adding a coagulant aid. The test differs from the preceding procedures in that the coagulant dose is varied, while the previously selected coagulant aid dose is held constant. The range of doses tested will be determined by the dose presently used in the plant. For example, in a jar test to evaluate a current dose of 100 mg/L alum under highly turbid conditions, a good range of doses to try might be 10, 30, 40, 60, 80, and 100 mg/L alum. In this example, the first jar would be dosed with 10 mg/L, with increasing doses in subsequent jars until 100 mg/L is reached.

If lime or another pH-controlling chemical is presently employed in the plant, this dose must be adjusted since its addition is meant to counteract pH changes due to coagulant addition. The amount added to achieve a particular pH should be adjusted to be proportional to the coagulant dose. The titration procedure discussed earlier will provide the needed data.

In addition, the final pH of each jar test should be measured. If these pH values are not within 0.5 pH units of the settled water in the plant, readjust the lime doses and repeat the jar test.

A more time-consuming procedure can be used to locate the precise combination of optimal coagulant and coagulant aid doses. An entire set of coagulant aid additions can be evaluated at each primary coagulant dose, thus covering many combinations of dosages. For example, six polymer doses (perhaps 0, 0.1, 0.25, 0.5, 1.0, and 1.5 mg/L) could be employed in each jar test set, while maintaining a constant alum dose. The procedure would be repeated using alum doses (in each of the six jars) of 10, 25, 50, 75, and 100 mg/L alum. The optimal dose combination can then be determined. The disadvantage of this approach is obviously the large number of jar tests that must be run; note that a very large raw water sample should be obtained so that all of these tests are run on the same water. This precaution is particularly important if coagulation is being evaluated under storm runoff or other rapidly changing conditions.

Nonconventional Treatment

Attempts are sometimes made to apply jar testing procedures to nonconventional treatment processes such as upflow clarifiers, sludge blanket clarifiers, and softening. No jar testing procedures have been established for nonconventional settling basins such as upflow clarifiers and sludge blanket clarifiers. Some manufacturers have applied empirical methods to develop procedures for evaluating their process units, but some of these procedures may be of questionable value.

One manufacturer of an upflow sludge blanket clarifier recommends that a "stickiness coefficient" be calculated with the aid of a special test apparatus to which "jigs" of chemically treated water are added to simulate the pulsing of the sludge blanket. The applicability of this type of bench-scale procedure to full-scale treatment units has not been established.

Some nonconventional processes allow reasonable empirically established modifications to the standard bench-testing procedures. For example, one bench-scale simulation of a lime softening process including a solids contact unit required the addition of 2.5 g/L of calcium carbonate to the jars during the flocculation step in order to generate results similar to those from the full-scale facility.

REFERENCES

AH Environmental Consultants, Inc. 1997. Disinfectant/Disinfection By-Product Rule at the Naval Station Roosevelt Roads, PR. Final Report to Naval Facilities Engineering Command, Atlantic Division. Hampton, Va. August.

AH Environmental Consultants, Inc. 1998. Evaluation of Alternatives for Enhanced Coagulation and DBP Control for the Goldsboro Water Treatment Plant. Final Report to Department of Public Utilities, City of Goldsboro, N.C. Hampton, Va. January.

City of Phoenix, Arizona. 1989. *Water Quality Master Plan.*

Cornwell, D.A., and M.M. Bishop. 1983. Determining Velocity Gradients in Laboratory and Full-Scale Systems. *Jour. AWWA,* 75:9:470–475.

Dentel, S.K., et al. 1986. *Procedures Manual for Selection of Coagulant, Filtration, and Sludge Conditioning Aids in Water Treatment.* Denver, Colo.: American Water Works Association and AWWA Research Foundation.

Environmental Engineering & Technology, Inc. 1987. A.B. Jewell Laboratory Treatability Study: City of Tulsa, Okla. Newport News, Va. September.

Environmental Science & Engineering, Inc. 1981. City of Arlington, Tex., Water Treatment Plant Process Upgrade and Trihalomethane Reduction Study. Final Report. ESE No. 81-209-200. Gainesville, Fla. December.

———. 1982. City of Arlington, Tex., Southwest Water Treatment Plant. Process Design Study. ESE No. 82-204-400. Gainesville, Fla. June.

Fair, G.M., J.C. Geyer, and D.A. Okun. 1968. *Water and Wastewater Engineering.* Vol 2. New York: John Wiley & Sons.

Hannah, S.A., J.M. Cohen, and G.C. Roebeck. 1985. Measurement of Floc Strength by Particle Counting. In *Drinking Water Supply and Treatment, Selected Papers by Gordon C. Roebeck.* Denver, Colo.: American Water Works Association and AWWA Research Foundation.

Hudson, H.E., Jr. 1981. Jar Testing and Utilization of Jar Test Data. In *Water Clarification Processes: Practical Design and Evaluation.* New York: Van Nostrand Reinhold.

Scott, J.S., and P.G. Smith. 1980. *Dictionary of Waste and Water Treatment.* London: Butterworth.

Singley, J.E. 1981. Coagulation Control Using Jar Tests. *Proc. AWWA Annual Conference-Coagulation and Filtration: Back to Basics*. St. Louis, Mo.

American Public Health Association, American Water Works Association, and Water Environment Federation. 1998. *Standard Methods for the Examination of Water and Wastewater*, 20th ed. Washington, D.C.: APHA.

Chapter **2**

Streaming Current Detectors

Water treatment plant operators have long searched for simple, reliable, and inexpensive equipment to assist them in selecting the proper coagulants and the most cost-effective dosages. Underdosing or overdosing of coagulant chemicals often occurs, especially during times of rapidly changing water conditions. Often, a water treatment problem is discovered after the water is in the clearwell.

The impact of the Surface Water Treatment Rule in the late 1980s compounded these problems by increasing the number of filtration plants and establishing more stringent water quality criteria. More recent regulations regarding enhanced coagulation (under the Disinfectants/Disinfection By-products Rule) and microbial removal (under the Enhanced Surface Water Treatment Rule) have further demonstrated the need to optimize coagulation processes for removal of particulate matter and organics. This group of regulations will expand the importance of effective coagulant control.

One instrument for coagulant control that has gained acceptance among plant operators and engineers is the streaming current detector (SCD). The SCD measures net colloidal and coagulant surface charges of a water sample following chemical addition. The instrument tells the operator, immediately after chemical mixing, whether the proper quantities of chemicals have been added. This timely information allows a change (either manual or automatic) in coagulant dose before water reaches the filters.

This chapter discusses the theory and background of the SCD. It also summarizes some of the general history and experiences with this equipment and indicates expected costs and benefits for water treatment facilities. Much of the information in this chapter is summarized from a report, *An Evaluation of Streaming Current Detectors,* prepared by Steven K. Dentel and Christine M. Kingery for the AWWA Research Foundation (Dentel & Kingery, 1988). The 1988 report is a good reference for anyone involved in the design or operation of a treatment plant using an SCD. This chapter also reflects some of the changes in technology and equipment use since the 1988 report. Reports of current manufacturing data and cost information came from two suppliers of SCDs (Chemtrac Systems, Inc., Norcross, Ga., and Milton Roy Company, Ivyland, Pa.). Reports of actual operating experience were provided by the American Waterworks Service Company and many water treatment plants across the United States (Newman, 1989 and 1995). Particular emphasis is placed on the overall experience of the Pennsylvania American Water Company.

BACKGROUND

The SCD was introduced by Gerdes in the mid-1960s (Newman, 1989). Gerdes (1966) described the unit as "a plastic boot containing a dead-ended bore, electrodes, and loosely fitted piston which forces the liquid back and forth through the annular space between the boot and piston." Experiments by Smith and Somerset (1969) demonstrated that an electrokinetic phenomenon other than zeta potential could accurately predict the behavior of a colloid–coagulant system. The SCD was shown to be capable of monitoring the subtle ionic and colloidal interactions that influence the efficiency of the coagulant in the removal of colloidal matter from source water or wastewater.

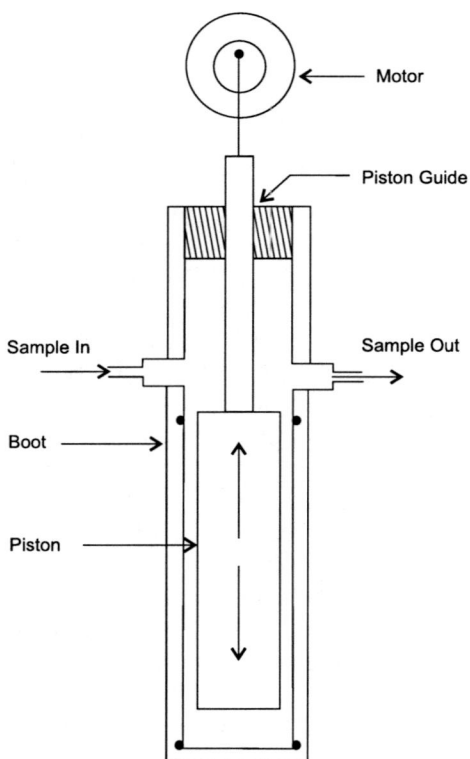

Note: Some types can be used in batch mode (breaker sample) as well as the indicated flow-through configuration.

Source: Source: Denatel and Kingery (1988).

Figure 2-1   Sensor of the streaming current detector

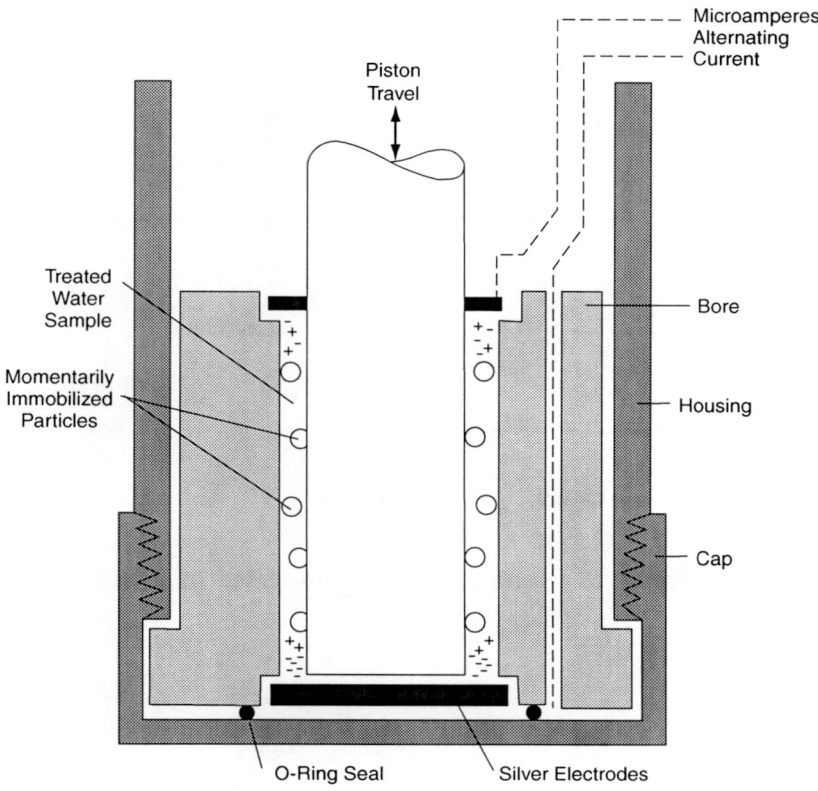

Source: Dental and Kingery (1988).

Figure 2-2   Electrode locations in the streaming current detector sensor

A simplified cross section of an SCD sensor chamber is presented in Figure 2-1. A sample of water flows through the chamber, typically at a flow rate between 0.25 and 0.75 gpm (1 and 3 L/min), where a small piston moves in a vertically reciprocating motion. A current is generated as electrically charged particle momentarily attach to the piston and the inner-boot surfaces. As shown in Figure 2-2, electrodes in the cylinder measure the current, and electronic processing generates an output known as the *streaming current*. Although the value is given in relative units, a standardization procedure can be used to provide consistent readings that are more easily related to zeta potential or electrophoretic mobility (see Chapter 4).

APPLICATIONS AND BENEFITS

The SCD was used in paper mills, wastewater treatment plants, and other chemical treatment applications that incorporate coagulation in their treatment processes long before it appeared in the water industry. The use of SCDs in water treatment plants began to increase in the early 1980s, and its popularity has continued to expand in the 1990s. The typical placement of an SCD in a water treatment process is shown in Figure 2-3. The streaming current level at which optimum coagulation occurs is dependent on and should be determined for each

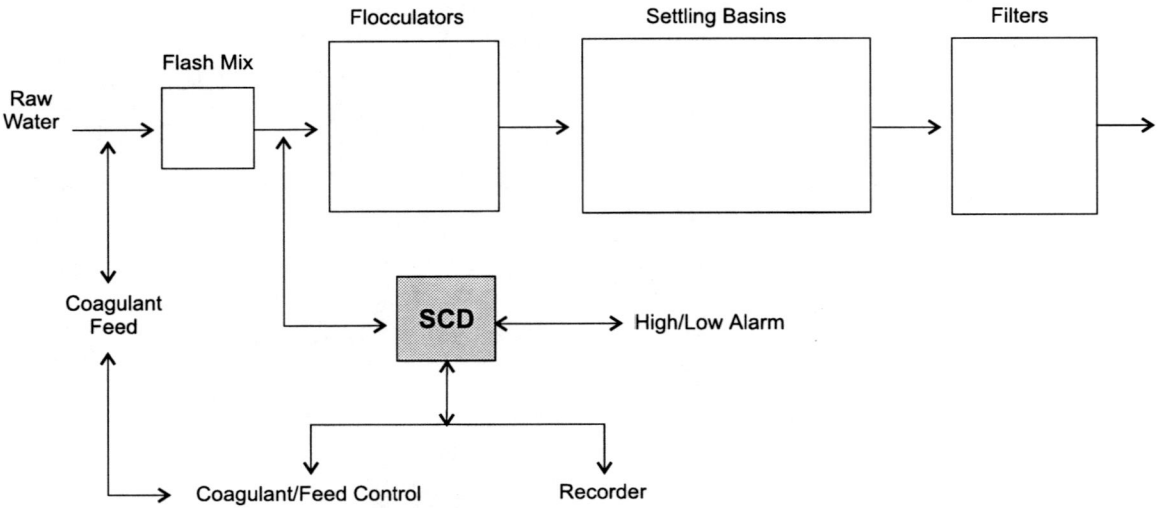

Source: Dentel and Kingery (1988).

Figure 2-3   Typical placement of streaming current detector in a process sequence for water treatment

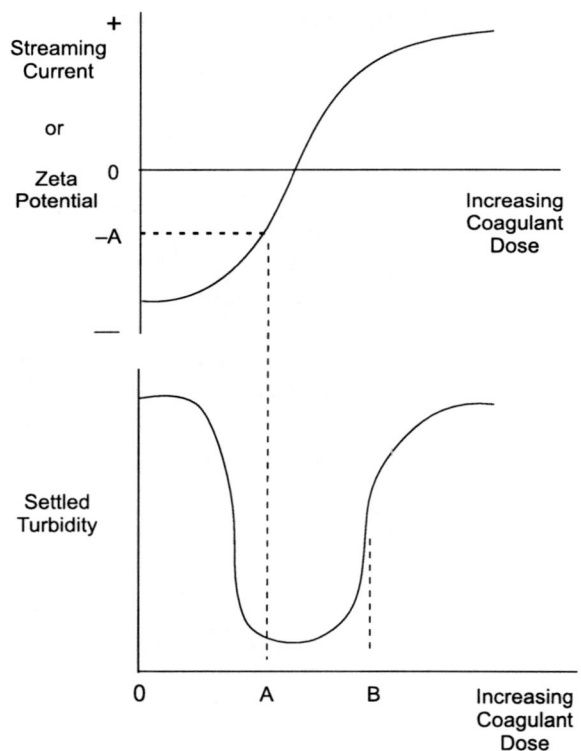

Note: Particle charge may be indicated by either zeta potential or streaming current. Types of behavior other than that shown are also possible, especially at high doses of inorganic coagulants. In this example, the setpoint might be set a dose "A."

Source: Dentel and Kingery (1988).

Figure 2-4   Simplified relationship between particle charge and final turbidity: Example 1

particular raw water source and process. A simplified relationship between particle charge and treated water quality is presented in Figure 2-4.

In the early 1980s, when SCDs began to find widespread use in water treatment plants, a number of benefits in monitoring and controlling coagulant feed rates were reported:

- Better control over coagulant dosages, especially under variable water quality and flow conditions

- Reduced water treatment cost, primarily due to more efficient adjustments of chemical applications when raw water quality varied

- Reduced process residuals due to better dose control

- Improvement in net water production through lengthening filter runs

- More consistent finished water quality

- Prevention of plant upsets and early detection of equipment malfunctions

MANUFACTURERS

Currently, two companies are major suppliers of SCDs in the United States, with others in Europe. Several hundred units are in use for water treatment applications. The domestic suppliers are the Milton Roy Company and Chemtrac Systems.

The reported installed cost (1996 dollars) of an SCD and monitor range from $8,000 to $10,000. An SCD system with chemical control capabilities is reported to cost from $10,000 to $13,000, depending on requirements specific to each water treatment plant. In larger water treatment plants or plants with variable source water quality, the SCD is shown to be a cost-effective tool, as discussed in more detail later in this chapter.

EVALUATION OF STREAMING CURRENT DETECTORS

An AWWA Research Foundation report (Dentel & Kingery, 1988) provided the first broad assessment of SCD usage in water treatment facilities. In part, this project evaluated possible benefits of SCDs and their potential roles in workable control techniques for chemical and dose selection in water treatment plants. The three primary goals of the project were to:

1. Determine the SCD's mechanisms of comparing streaming current with other coagulation parameters, such as zeta potential and turbidity, and assess any functional problems with the SCD via laboratory and theoretical studies.

2. Provide a database with input from as many plants as possible with experience using the SCD via a survey of users.

3. Closely analyze SCD applications, including comparisons of water quality and operating experience before and after SCD installation, via on-site studies of SCD use at a small number of water treatment plants.

This chapter summarizes the results of the study and, in some respects, updates them. The original AWWARF report provides additional information with regard to theoretical aspects of the SCD measurement technique and practical aspects of SCD installation.

Table 2-1   Factors influencing SCD performance

Factor	Effect	References
Temperature	Slight to moderate increase in sensitivity with temperature, dependent in part on materials used for sensor	Gerdes (1966), Galetti (1969)
Response time	Response time varies from instantaneous to several minutes, longer for complete stabilization, particularly for desorption of surface active materials	Gerdes (1966), Galetti (1969), Dentel and Kingery (1988)
Soluble salts	Higher concentrations cause decreases in SCD reading, though usable response observed up to 0.1 percent salt; effect varies by salt; specific adsorption a factor for multivalent ions	Gerdes (1966), Galetti (1969), Dentel and Kingery (1988)
pH	Similar to effect on zeta potential: pH increase gives more negative values for simple systems, but trends are more complex following coagulation with alum or ferric salts; potential loss of sensitivity at pH > 8	Galetti (1969), Dentel and Kingery (1988), Dentel (1995)
Flow rate	More rapid sample flow decreases time required to equilibrate	Galetti (1969)
Particle size	Responds to materials in both colloidal and subcolloidal ranges, including surfactant and humic substances	Dentel and Kingery (1988), Dentel (1995)

Factors Affecting SCD Operation

Table 2-1 summarizes some of the factors that influence SCD operation. Some of these factors result from aspects of the coagulation process itself, and others from the way that the SCD functions.

Chemical factors.   Table 2-1 shows that two chemical properties of the water (ionic strength and pH) should be taken into account when interpreting SCD readings. These variables do not generally affect routine use of the SCD, but they may be important in specific applications. Generally, ionic strength is sufficiently constant in water treatment to ensure negligible effects on the SCD reading. Facilities experiencing considerable fluctuations in ionic strength or conductivity of a raw water (due, for example, to tidal effects on estuarine water sources) have seen variability in SCD readings; some of this phenomenon correctly reflects the effects of double layer compression on particle stability, but higher conductivity levels (greater than 1,000 microsiemens/cm) can reduce efficiency of the SCD instrument by reducing its sensitivity.

The pH can affect SCD readings in two situations:

1. At almost any pH, the surface charge and zeta potential of coagulated particles will change if the pH is altered. The SCD reading will also change in a similar fashion. This behavior can be confusing to operators if the coagulant dose has not been changed. However, the SCD reading in such a case is a *correct* indicator of coagulation effectiveness, as the zeta potential (and not pH) is the more direct indicator of particles' tendency to coagulate.

2. High pH tends to accompany negative charge in both the SCD sensor surfaces and aqueous particles. Some evidence suggests that this combination leads to inefficient particle coverage of the SCD sensor and lack of instrument sensitivity, particularly when the particle concentration is low.

The problem has only been observed with specific waters and coagulants above pH 8, and it is unlikely to arise in most treatment applications.

Response time and lag time.   Time is another factor of considerable importance in proper use of the SCD. Table 2-1 indicates that the instrument's response time may be up to several minutes if an immediate change in conditions occurs. If the SCD is used for feedback control of the coagulant dose, this period also leads to a delay in coagulant feed adjustment. However, such a delay is unlikely to cause problems in actual treatment, because most SCDs respond quite rapidly to changes; only the final adjustment to a precise level usually requires additional time. The dose adjustment will thus follow a similar pattern. In addition, changes in water conditions are usually gradual in nature rather than instantaneous, and mixing effects in processes downstream from coagulant addition will mitigate any momentary deviations in conditions.

In addition to the response time, another important consideration is lag time, which is the elapsed time between coagulant injection into the raw water and the time when that same water contacts the SCD sensor. This period precedes the actual instrument response time, but the two must be added together to determine the overall delay in response by an SCD to a change in coagulant dose. Furthermore, the lag time is important, because the coagulation process must be sampled after an appropriate reaction time.

When a coagulant is added to the water, it combines with the colloidal material, which initially is negatively charged. In theory, the coagulant neutralizes this charge to give neutrally charged (uncharged) particles, which then flocculate with no charge repulsion.

In reality, however, this sequence is a dynamic process. Chemical reactions achieve charge neutralization over time, so a neutral streaming current would be measured slightly downstream from the point of coagulant addition. If an SCD is used to sample the flow even farther downstream, the reading may gradually revert to a more negative value. This effect may result from a variety of factors, but the practical implication is that the lag time affects the SCD reading. Therefore, the SCD sampling location and lag time must be carefully selected.

A rule of thumb calls for a 2-min to 5-min lag time between the coagulant addition and monitoring by the SCD. This period may vary depending on specific plant conditions. The lag time can be intentionally changed in two ways: (1) by selection of the sampling point (where the sampling line draws from the process flow), or (2) by altering the flow time required for the sample to reach the SCD.

Location of sampling point.   The sampling point should follow thorough, rapid mixing to assure uniform dispersion and reaction of coagulant. Rapid fluctuations in the SCD reading (particularly during extremes in process flow) may indicate nonuniform dispersion of coagulants and a need to sample further downstream.

The sample point should *not* be located near a wall or low point in a basin, channel, or pipe, because ineffective mixing may occur at these spots, leading to inadequate instrument response or signal noise. Sand, grit, or other abrasive materials sometimes found at these locations could damage the sensor or clog sample lines. Sampling near the center of a pipe, through a corporation stop diffuser, has been a successful strategy. If possible, avoid pumping the sample flow, because this configuration introduces the possibility of pump failure.

Each treatment facility is a unique case, with complications such as limited availability of sampling locations and addition of chemicals at multiple points. When determining the point at which to sample, *consult the SCD manufacturer* for guidelines based on experience at similar facilities. It is also advisable to evaluate different sampling points in order to assure maximum instrument response with a minimum of signal fluctuation. Perform this evaluation under a variety of anticipated conditions. Use of a chart recorder or other graphical display of trends in the SCD readings will aid in this task (and also in selection of the SCD's set point for optimum coagulant dosage).

Obtaining the proper sample flow time.   The lag time between coagulant addition and streaming current measurement has two components: (1) the time within the process before the sampling point, and (2) the flow time within the sampling line. Ideally, time within the process is optimized by proper sample point selection. The flow time should then be minimized in order to keep the overall lag time at a minimum. A long lag time between chemical addition and the SCD allows significant delay in the response to changing conditions, reducing the benefits of the SCD. Proper and timely response is especially important in SCD systems that automatically control chemical dosages.

Flow time can be minimized in several ways. Most important is placement of the SCD (or at least the sensor component) as close to the sampling point as possible, in order to limit the length of the sampling line.

For a circular pipe or tubing, the flow time can be calculated by:

$$\text{Flow time} = \frac{3.14 \times \text{Length} \times \text{Diameter}^2}{4 \times \text{Flow rate}} \qquad \text{(Eq 2-1)}$$

This formula suggests that the flow time can also be decreased by using a smaller diameter sampling line. However, this change will also decrease the sample flow rate somewhat. The flow rate can be increased by boosting the elevation difference between the SCD sensor and the height of the sampled flow. If the flow time is decreased such that the overall lag time is less than the recommended 2 to 5 min, the period allowed for coagulation reactions may be inadequate to allow full effects to develop.

A time lag in excess of several minutes can be used if necessary, but if the system includes automatic dose control, the controller should be set to compensate for this lag. Otherwise, the detector/control equipment will be continually adjusting chemicals based on signals derived from "past" water quality rather than quality at the actual time of monitoring. "Cycling" of the streaming current measurement and chemical dosages might result, with values oscillating above and below the optimum level.

Theoretically, the optimum lag time may also be affected by variation of the treatment flow, which changes the time between coagulant addition and sampling. Variability in the time required for the coagulation reactions may also result due to such factors as temperature and coagulant dose. Changes in the streaming current value for optimum coagulation are not likely to be significant, but appropriate adjustment of the set point can compensate if required.

In summary, SCD technology can succeed only with proper installation and sample point selection. Manufacturers should be consulted during the planning process to ensure correct setup.

Factors Affecting Dose

Dentel and Kingery (1988) identified various factors that influence coagulation and coagulant dose. Many are accommodated in the SCD reading. The following factors are the most important ones.

Flow rate. A change in water plant flow rate must be met by a proportional change in chemical feed rate to maintain the proper coagulant dosages. Use of the SCD for dose control allows automatic compensation for such changes in flow. The SCD will indicate an increasing or decreasing plant flow rate by a corresponding change in the streaming current, and this indication allows for an increase or decrease (manual or automatic) in chemical feed rates to maintain the optimum streaming current level.

If desired, an automatic dose control system can also be set to adjust dose primarily on flow (based on input from a flow measurement device) with a trimming signal sent by the SCD. In this case, the SCD is used only to fine-tune the dose based on water and coagulant characteristics. The advantages of this control method are rapid response to flow changes (through feedforward rather than feedback control) and separation of control functions. The disadvantage is a more complex control system that may be more difficult to troubleshoot.

Coagulant strength. Infrequently, the strength or type of coagulant chemicals used may vary. The SCD will then detect changes in the streaming current level. Such a signal allows for either the automatic or manual adjustment of chemical feed rates to compensate for changes in coagulant strength.

Solids concentrations. An increase in particle concentrations (turbidity) or color-causing organic matter often increases coagulant demand. By maintaining an optimum streaming current level through changing chemical dosages, a system can ensure optimum treatment efficiency despite varying source water quality conditions. Under rapid and excessive changes in source water conditions, an SCD can be of the most benefit to the water treatment plant.

pH. Along with other factors, pH has been shown to have a strong effect on the streaming current. When an inorganic coagulant is used, the coagulation pH is shown to affect floc volume, with some influence on the process of flocculation. In general, this relationship will not interfere with use of streaming current as an indicator to monitor proper coagulant dosages. To reduce interference between coagulants and pH control chemicals, both sets of chemicals should be dosed in a constant proportion (based on the streaming current) with the ability to fine-tune this proportion when necessary. Particular difficulty has been reported where ferric coagulants are used in conjunction with pH control agents.

Particle charge. The coagulant dose required to reach a predetermined SCD set point sometimes changes over time, even with constant source water turbidity. This effect may be due to variation in colloid properties such as size or charge density. For example, a rise in the molecular weight of NOM increases the coagulant requirement without affecting measured turbidity. However, the SCD will detect the influence of NOM on particle charge following coagulation, so it is likely to be appropriate even with enhanced coagulation.

Signal offset. As indicated in Table 2-1 and the preceding discussion, the SCD reading corresponding with the desired coagulation efficiency (that is, the set point) is not necessarily a value of zero, despite expectations based on theory. Causes may include incomplete sensor coverage by colloidal matter or inability to locate or access the optimum sampling point. An unavoidably long sample flow time may have the same result. The effect is actually inconsequential in practice. As long as the relative changes in the SCD reading are consistent, the set point can be used successfully, whether it is exactly at zero or not.

Other factors. The optimum streaming current set point sometimes varies either monthly, seasonally, or in some cases daily, based on source water conditions. A number of benefits of SCD technology have been identified in such cases. The SCD can indicate the effects on sludge blanket clarifiers of inconsistent feed rates, improper rapid and flocculation mixing, recycling backwash water, occasional use of powdered activated carbon, and varying ratios of different coagulants.

INSTALLATION AND OPERATIONAL DIFFICULTIES

Chapter 4 from Dentel and Kingery (1988) presents potential difficulties with the installation and operation of SCDs. Reported problems have included clogging, abrasion from natural solids of added treatment chemicals (lime, PAC), coating of sensor surfaces, and inadequate documentation furnished with the instruments. The most common problem appears to be clogging of sample lines. Remedies to this problem include:

- Relocation of sample points to reduce entry of silt or other material into the line

- Use of sample lines made of transparent material and observable flow drains, so that flow can be checked during regular operator rounds

- Use of small settling units (or other separators) to remove larger solids in sample lines prior to the monitor (Filters provide satisfactory removal, but they may be impractical if frequent clogging of their screens or media blocks flow to the SCD.)

- Periodic cleaning of lines with pressurized air or backwash water

- Increased flow rates through the unit

- Short sample lines without mad traps

Another serious problem is grit in the sample, which can reduce or destroy probe sensitivity. Solutions to this problem are similar to those for clogged sample lines.

Installation and Operational Difficulties Summary

This section gives a summary of findings and conclusions of the Dentel and Kingery (1988) research project.

1. The SCD provides information similar to measures of zeta potential. Laboratory use of the instrument may be more difficult than plant use because of background problems, signal offset, and an arbitrary amplification factor. However, useful results may be obtained with proper attention to these factors.

2. The SCD reading correlates with turbidity removal, especially when charge neutralization is the dominant coagulation mechanism. A signal offset may be observed, but it has no harmful impact on use of the SCD system.

3. Treatment plant users of SCDs are generally satisfied with the instrument. It provides information more rapidly than a jar test does. More important, under most conditions it also shows in which direction the coagulant dose should be adjusted.

4. According to plant records, use of SCDs cut chemical use by an average of 12 percent under stable source water conditions and 23 percent under changing conditions. Savings are more substantial under changing conditions (such as storm or high runoff events), which helps to offset the cost of an SCD.

5. An SCD is also useful for indicating pump failures, rapid mix failures, feed line clogging or damage, and operator inattention, as well as for comparison of parallel processes, chemical feed points, etc.

6. Proper functioning of an SCD depends on proper installation, particularly the location of the sample point. A sample flow that is free of abrasive grit and resistant to clogging must be assured. Moving the sample point downstream protects against these problems but increases the lag time between dose and response.

7. Like any other piece of equipment, an SCD requires scheduled preventive maintenance. If the instrument is used for automatic control of the coagulant dose, this is a particularly important precaution. Cleaning may be required from every 2 days to every 6 months. Source water that is gritty or high in iron or manganese is more likely to cause problems. Polymer coagulants may be easier to accommodate than inorganics.

8. Prior to purchasing an SCD system, a trial period is strongly recommended. The evaluation period should include conditions when coagulation difficulties are encountered, and it should last long enough for a valid assessment to be made. Installation assistance should be included with the sales agreement. The purchase should be backed by a performance agreement, where possible.

9. SCD technology is not a substitute for good operation or management. In fact, effective practices are prerequisites for optimal use of the instrument. Periodic comparison with jar tests or zeta potential is recommended, especially before plant personnel have accumulated experience with the SCD under all operating conditions. In addition, the SCD is not effective in polymer feed optimization.

A report by Cleasby et al. (1989) found that the SCD was a useful tool in many plants, achieving good performance at high filtration rates.

FIELD APPLICATIONS AND EXPERIENCES

The American Water System is a private water purveyor encompassing numerous water treatment plants located throughout the United States. In the early 1980s, the American Water System began to investigate the potential uses of streaming current technology for process control. Streaming current detectors/monitors were tested at various sites under varying raw water qualities. Treatment systems dealing with low hardness/low alkalinity supplies in the northeastern United States were evaluated, along with plants on major industrial waterways in the Midwest. Comparisons with zeta potential, jar tests, and coagulation control centers utilizing pilot filters were conducted versus streaming current with favorable results. The term *streaming current detector* refers to all machines based on the streaming current principal, including equipment specified as *monitor*.

A survey was conducted in 1995 on the use of streaming current detectors in the American Water System by a focal group evaluating the reliability of equipment for plant automation. Questions posed to plant personnel in the survey were aimed at the installation of the SCD, mixing time/energy, time from coagulant addition to sample analysis by the SCD, training, maintenance, and overall uses of the SCD for process control.

The survey showed that cases of unsuccessful SCD use could be tied to several factors. Several plants reported that the SCDs did not perform well, but further investigation revealed overly long or short sample times from coagulant addition. In plants where the SCD was being used successfully for coagulation control information, sample times were set between 3 and 5 min, which is generally recommended from the various manufacturers.

Seven out of the 26 plants utilizing the SCD were set up in a coagulation control mode which enabled the SCD to pace the chemical feed pump. Typically, dosage control was primarily paced off raw water flow and fine-tuned via SCD signal (a compound-looped configuration). Alum was the primary coagulant at most of these sites, with very infrequent or no pH preadjustment chemical feed.

One system in Hopewell, Va., was using an SCD to measure the tidal influence on its source water. Quality changes typically occurred every 12 h. The system maintained a constant delta value between the source water and treated water potential as these tidal influences occurred, resulting in optimal coagulation at all times under varying source water quality conditions.

Approximately one half of the respondents reported making seasonal adjustments to targeted SCD values as water temperatures changed. The SCD was proven to be successful in maintaining proper coagulant feeds under both warm-water and winter conditions.

Maintenance of the units was very intensive in plants utilizing treatment chemicals containing iron salts and routine additions of powdered activated carbon and/or potassium permanganate. Plants utilizing sources containing high levels of iron or manganese also required high maintenance. Daily cleanings were reported at some sites with these conditions present. The overall average cleaning frequency needed to maintain proper operation was determined to be once per week.

Uses of Streaming Current Detectors in Pennsylvania

The Pennsylvania American Water Company (PAWC), an operating subsidiary of the American Water System, began to investigate the use of streaming current technology for coagulation control in the early 1980s. One unit was purchased, mounted on a stand, and transported from plant to plant for evaluation. Early results indicated a probability of successful implementation if proper installation procedures were followed, ample training were offered to all operating personnel, and maintenance of the unit were conducted at regular intervals. The streaming current detector could effectively alert plant personnel of loss or diminished coagulant feed due to feed pump failure, in addition to its role as a tool for plant operators to maintain optimized coagulant addition at all times.

As of early 1996, PAWC had outfitted 17 water treatment plants with SCDs. Of these systems, three configured the SCDs in compound-looped arrangements where coagulant addition was based primarily on source water flow, with fine-tuning through the SCD. Problems were initially seen in one central PAWC plant site controlled only by SCD. Sluggish response to flow changes resulted in improper coagulant dosages. A change of the plant process control to compound-looped operation resulted in proper treatment under all conditions.

Recommendations for Successful SCD Operations

The SCD has proved to be very effective in coagulation control or monitoring in PAWC, if a number of procedures are followed:

1. Sample times from coagulant addition to SCD analysis must be kept at 3 to 5 min at average plant flow rates.

2. The SCD must be installed far enough downstream of the coagulant addition point to allow vigorous mixing. It is advisable to sample downstream of a rapid mix unit or in-line mixer, or after several pipe bends

if the stream undergoes no mixing. Adequate dispersion of coagulant into the center of the pipeline utilizing a diffusion corporation stop greatly assists in gathering a stable SCD signal.

3. Chemicals used to adjust pH (lime and caustic soda) are known to cause negative interference in the SCD signal. Changes in pH, either caused by chemical addition or naturally occurring in the source water, can affect SCD readings and lead to improper treatment. To avoid this problem, a configuration might collect samples for the SCD between the coagulant and the pH adjustment chemical feed points. This setup should be evaluated at the specific plant site prior to permanent installation.

4. Flow rates must be maintained within manufacturers' specifications under all normal source water pump flows. Changes in water flows through the units can cause changes in the SCD set points and may not reflect actual treatment conditions. A flowmeter and control valve should be installed on the drain line of the unit to allow adequate measurement and control of the flow.

5. Implementation requires initial plant personnel training and refresher sessions to ensure that the SCD signal is properly interpreted and the unit is properly maintained. The manufacturer should provide initial training upon installation.

6. Multiple operator selectable modes of operation should be provided, if control of the chemical feed pump is desired. As a minimum, the operator should be able to select a *manual* mode (in which feed pump output settings are adjusted by plant personnel), *flow-paced* mode (in which the coagulant feed pump is adjusted on flow only), or *compound-looped* mode (in which the coagulant feed pump output is adjusted primarily on process flow and fine-tuned based on the SCD signal). The manual and flow-paced options are useful when normal maintenance is needed on the SCD, or if problems are experienced with the pacing signal. These two settings also allow the SCD to be placed in the monitoring-only mode in order to arrive at the proper set point for operation.

Determination of the Proper Set Point

The process value is determined by first optimizing the plant particulate removal process and then noting the corresponding SCD value (set point). Factors which should be evaluated include filter effluent turbidity levels, filter run times, head loss buildups, time from filter to waste, and if possible, zeta potential and jar test information. A device, appropriately named a *controller*, continually measures the SCD reading and then adjusts the output of the coagulant control feed system. A modern SCD design allows plant personnel to set the arbitrary number on the readout to a value of zero. Overfeed conditions are then indicated as increasing positive readings on the SCD, while underfeed is documented as increasingly negative readings. The operator then knows immediately in what direction the coagulant dosage needs to be changed.

PAWC uses one main method for arriving at the proper set point. The SCD is first set up in a monitoring only mode (manual or flow-paced operation). Treatment conditions are then optimized for maximum plant process efficiency. Once operations are acceptable, the reading on the SCD is noted. This value then becomes the number which the operators strive to maintain under varying flow and turbidity conditions.

Once the set point is established, changes to the coagulant dose are performed manually to cause both underfeed and overfeed conditions. The SCD readout is then monitored as coagulant feed is adjusted away from optimal levels. When conditions begin to deteriorate in the plant operations (i.e., filter runs become shorter, filter effluent turbidity levels rise), the readings are recorded and the coagulant set back to maintain the normal set point. The readings gathered can then be entered into the controller as the range of operation of the automatic control. An alarm set up to be triggered by SCD readings outside these ranges can alert plant personnel of potential loss of treatment efficiency.

Plant Personnel Acceptance of the SCD

New instrumentation at a water treatment plant can create some skepticism by the operating staff, particularly if the system threatens the authority of shift operators to make decisions. An operator gains knowledge, trust, and experience in this technology if the instrument is first set in the monitor-only mode. The plant operator can then make adjustments to treatment as necessary, while still noting changes in SCD readings. Over the next couple of weeks, operators may become aware of increasingly negative SCD readings when the coagulant feed line plugs. Confidence rises as operators notice that increases in source water turbidity levels are adequately corrected by manual adjustments in coagulation doses, while SCD readings remain near those associated with higher turbidity conditions. Over time, confidence in the instrument is gained and acceptance for its use is developed. SCD control of the coagulant feed system can then be initiated using a compound-looped signal.

Maintenance of the SCD

As with all mechanical monitoring devices in a plant, a streaming current detector requires maintenance to assure the reliability of readings. Several factors powerfully affect the run times of these units. Some source water conditions, such as high turbidity and high iron and manganese content, tend to foul the electrode bore faster than others, leading to increasing frequency of cleaning. Chemical cleaning of oxidized iron and manganese from the electrode and unit body is best accomplished by the periodic addition of a reducing agent such as sodium oxalate or sodium bisulfite.

Sample and drain line plugging is a frequent problem at sites with naturally high volumes of grit in their source water supplies or in systems that feed powdered activated carbon, bentonite clays, or lime prior to the SCD sampling point. Cleaning frequencies must be determined at each plant site and proactive maintenance scheduled.

SUMMARY

Since the early 1980s, the streaming current detector has gained widespread use in the water treatment industry. The use of SCDs on a variety of raw water supplies in various parts of the United States has demonstrated their ability to control chemical dosages, especially during variable water quality conditions. Potential for reducing chemical costs has been demonstrated due to increases in efficiency of chemical applications. More important, consistent water quality can be maintained. Additional benefits are possible, because the SCD can notify plant operators of potential upsets in plant equipment or efficiency in the various process units. Proper installation and maintenance of the SCD has been shown to provide much improvement in the quality of water leaving a treatment facility. For large plants or those experiencing great variations of water quality, the unit has shown to be a cost-effective investment with payback occurring in less than 5 years.

REFERENCES

Cleasby, J.L. et al. 1989. *Design And Operation Guidelines For Optimization of the High-rate Filtration Process: Plant Survey Results*. Denver, Colo.: American Water Works Association and AWWA Research Foundation.

Dentel, S.K. 1995. Use of the Streaming Current Detector in Coagulation Monitoring and Control. *J. Water SRT-Aqua*, 44:2:70.

Dentel, S.K., and K.M. Kingery. 1988. *An Evaluation of Streaming Current Detectors*. Denver, Colo.: American Water Works Association and AWWA Research Foundation.

Galetti, B.J. 1969. Application of the Streaming Current Detector to the Continuous Measurement and Control of Colloidal Systems. 15th National ISA Analysis Instrument Symposium.

Gerdes, W.F. 1966. A New Instrument—The Streaming Current Detector. 12th National ISA Analysis Instrument Symposium, Houston, Texas.

Neuman, W.E. 1989. Streaming Current Detectors for Process Control. AWWSC Experience Paper. American Water Works Service Company, Voorhees, N.J.

Smith, C.V., Jr., and I.J. Somerset. 1969. Streaming Current Techniques for Optimum Coagulant Dose. 15th National ISA Analysis Instrument Symposium.

This page intentionally blank.

Chapter **3**

Particle Counting and Sizing

IMPORTANCE OF PARTICULATE REMOVAL

The removal of particulate matter is one of the most important goals of water treatment. Determining the number of particles in the raw water, what happens to the particles during treatment, and how many particles remain in the effluent are all critical steps in optimizing water treatment.

Particulate matter gives water an aesthetically unpleasant, cloudy appearance. In addition, many contaminants of concern to the water industry are either particles or associated with particles. For example, clays and silts, colloidal metals, and biological contaminants, including bacteria, viruses, and cysts, are all particulate matter. Furthermore, particulate matter has been shown to shield organisms from disinfectants. Berman, Rice, and Hoff (1988) reported up to 20- to 50-fold increases in the time required for inactivation of particle-associated coliforms depending not only on type of disinfectant but also on particle size.

Waterborne disease outbreaks have been associated with drinking water that had gone through conventional treatment plants, focusing attention on the need for more sensitive monitoring of filtered water. Currently, the only regulations dealing with particulate matter in drinking water are contained in the Surface Water Treatment Rule (SWTR). That rule requires that any facility using a surface source water treated by coagulation and filtration produce an effluent turbidity of 0.5 ntu or less 95 percent of the time. The SWTR is expected to be revised soon, and the turbidity standard will most likely be lowered. The absolute number of particles, as measured by a particle counter is not yet regulated. Some states have allowed the use of particle counting, however, to demonstrate the effectiveness of unconventional treatment methods such as high-rate filtration.

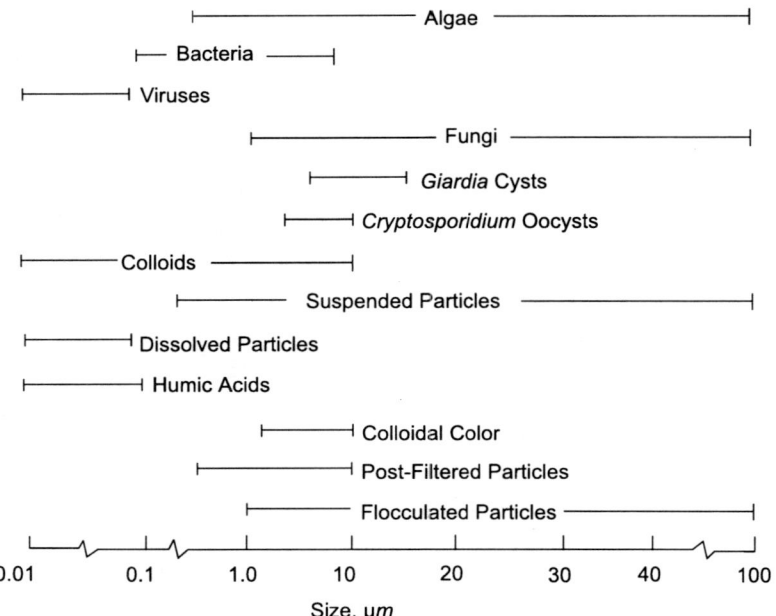

Source: McTigue and Cornwell (1988).

Figure 3-1 Particulates present in source and finished water

Figure 3-1 shows some of the substances and particles present in raw and finished water. Bacteria generally measure 0.1 to 12 µm, viruses are smaller than that, and algae can range anywhere from 0.5 to 150 µm. The human eye can begin to see particles of about 70 µm size. *Giardia* cysts are about 5 to 15 µm in size, and *Cryptosporidium* oocysts are about 4 to 7 µm.

A well-operated granular media filter can consistently remove particles 10 µm and larger, although filters can remove even smaller particles (Tate and Trussell, 1978). A flocculated particle may range in size from 1 µm up to 100 µm. A floc with particles of about 15 to 30 µm is considered to be optimal for settling.

Traditionally, measurements of particulate matter have been taken by evaluating either turbidity or suspended solids concentrations. Both are indirect measurements. Turbidity is a measure of the light-scattering properties of the particulate matter present in water. A light beam passes through a water sample, and suspended matter scatters a portion of the light in proportion to the matter present. The scattered light activates a photoelectric detector that converts the light energy to an electrical signal. Because the light-scattering properties of different kinds of particles vary, turbidity measurements cannot give direct data or particle concentrations. Suspended solids concentration is a gravimetric measurement of the mass of particulate matter present. If particle sizes or densities vary, suspended solids mass cannot be related to particle numbers.

Microscopic evaluation can provide correct readings, but it is a time-consuming process. The particle counter is the only technique available that can routinely provide a direct measurement of the size and number of particles present in a water sample.

Questions and concerns do exist about the accuracy and reproducibility of data derived from particle counters. In order to fully understand these problems and minimize them, it is important to understand how these instruments work, how they are calibrated, and how to interpret the data they provide.

Use of Particle Counters for Operational Control

Particle counters were developed for and are used extensively in industrial applications other than the water industry. The Coulter Counter (made by Coulter Electronics, Hialeah, Fla.) is used for blood plasma analysis, and similar light-blockage instruments are used in pharmaceutical, dairy, hydraulic fluid, and food-processing applications. Tate and Trussell (1978) first reported water quality data obtained from particle counting in the late 1970s, and the use of these instruments in the water industry is steadily increasing. As of 1997, more than 400 full-scale plants reported using particle counters for operational control. Many researchers report data obtained from particle counting in pilot-plant and laboratory-scale work.

Researchers suggest that particle counting can be used to optimize coagulation, sedimentation, filtration, and oxidation processes. Figure 3-2 shows a generalized example of the changes in number and sizes of particles that occur as a result of water treatment. Large numbers of many different-sized particles are present in a source water. The addition of chemicals and application of mixing energy during the rapid mix, flocculation, and sedimentation stages leave as many or more particles, but the particles are larger. Filtration then removes most of the particles, particularly those 10 µm and larger. Lawler, O'Melia, and Tobiason (1980) present an excellent discussion of the theories and mathematical relationships that can be used to predict how particle size and distributions change during water treatment.

A good example of the use of particle counters for control of coagulation and filtration is the experience of the Southern Nevada Water System, one of the first full-scale facilities to use particle counting to control coagulation. Experience in that system is described in detail later in this chapter. With a very low source water particle count, this system need not make dilutions, making it an ideal system for control by particle counting.

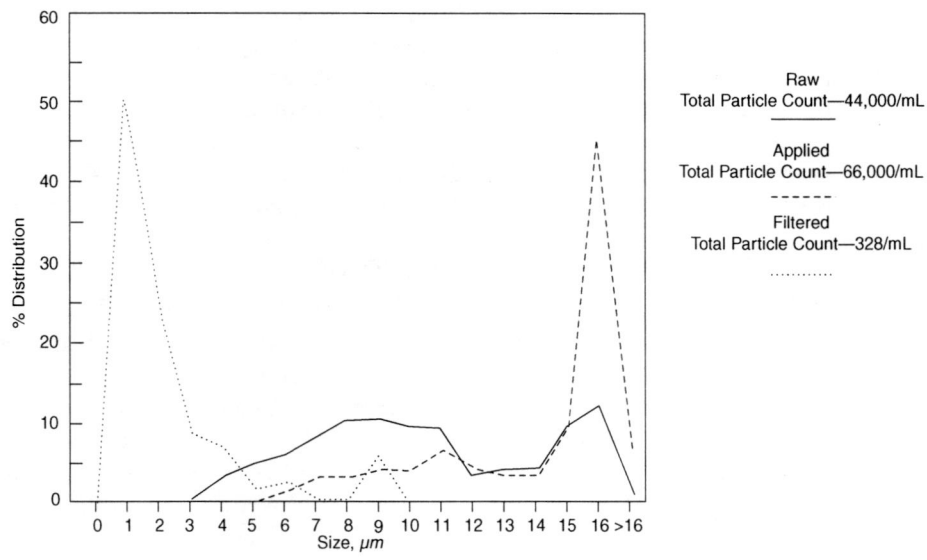

Detection Range = 0.5 µm to 10,000 µm

Source: McTigue and Cornwell (1988).

Figure 3-2 Particle distribution changes through treatment

Researchers have reported success in evaluating coagulation/flocculation processes on the bench and pilot scale with particle counting and sizing. Lawler, Izurieta, and Kao (1983) successfully used a Coulter Counter to study changes in particle sizes through laboratory treatment, particularly in flocculation. Researchers Saunier, Selleck, and Trussell (1983) have reported on the effects of ozonation on coagulation, using particle counters to demonstrate this effect.

TYPES OF PARTICLE COUNTERS

Microscopic Examination

A microscope is the only available instrument that allows a direct measurement of the particulate content of a water sample. Electron and optical microscopy have been employed as techniques to measure particle size distributions in drinking waters. Unlike the other particle counting techniques discussed in this chapter, only the use of an optical microscope allows the operator to detect and quantify particles in the size range of 1 µm to 1 mm. Transmission and scanning electron microscopes are used to measure particles smaller than 1 µm. They require sample evaporation and coating of the particulates with carbon or gold prior to analysis. Optical microscopy is the standard technique for quantifying planktonic particulates in natural waters. Membrane filtration prior to microscopic evaluation can be used to boost sensitivity and collect a permanent record of the sample. Generally, microscopic evaluations have been used for rough counting of planktonic and other particulate matter, such as asbestos fibers. The method has not been used to produce particle size distribution data, because two-dimensional microscopic images and other practical limitations prevent accumulation by visual counts of statistically adequate numbers of particles. Computerized microscopy can help to overcome these limitations.

Three different technologies are available in particle counting instrumentation—electrical sensing, light blockage, and light scattering. These three technologies are described in detail in the following sections. However, for full-scale applications, most plants use only light-blockage instrumentation. Laser light-blockage instruments can be used in online applications, the most useful type of this technology for water treatment plants. Instruments using electrical sensing and light scattering can still be used for water treatment, but because of the limitations presented by batch processing of samples, it is difficult to use them in daily operations of water plants.

Electrical Sensing Zone Method

To use the electrical sensing zone particle counting method, a water sample must be diluted with a mild electrolyte (typically NaCl) to increase the conductivity of the solution. This change may be achieved in natural water samples by adding an appropriate amount of more concentrated salt solution. Isotonic saline (0.9 percent) is used to avoid osmotic size changes in biological particles. The solution is pumped through a small orifice across which current is flowing. As a particle passes through the orifice, a voltage pulse proportional to the particle volume is produced due to the change in electrical resistance across the orifice. The pulses are electronically converted to enumerate and size the particles. Software supplied by the manufacturer generates a graphical presentation of the particle size distribution. Coulter Electronics first developed particle counters based on this principle and continues to refine its design. More recently, Particle Data, Inc. (Elmhurst, Ill.), has developed and marketed an instrument that is similar in principle to the Coulter unit. Figure 3-3 illustrates how a Particle Data instrument implements the "electrozone principle" to quantify and size particles.

For resistance-based models, the proportional relationship between the voltage pulse produced and particle volume is linear only for effective particle diameters of between 2 and 66 percent of the orifice. Therefore, equipment must use orifices of different sizes to cover a broad particle size spectrum. Typically, two orifices would be required to cover a particle size range of 1 to 100 μm. For example, the use of 30 and 300 μm orifice sizes would allow detection of a particle size range of 0.6 to 180 μm. Additional orifices may be used to achieve additional overlap.

Light-Obscuration or Light-Blockage Method

Hiac/Royco Instruments Division (Silver Spring, Md.), Hach (Loveland, Colo.), IBR (Grass Lake, Mich.), Particle Measurement Systems (Boulder, Colo.), and Met One (Grants Pass, Ore.), among other manufacturers, implement the light-obscuration technique in their particle counter designs. The sample is drawn through a channel under pressure, where it passes a window of known area through which a collimated light beam is passed at right angles to the fluid flow. As a particle passes through the light beam, its partial blockage of the light is picked up by a photodiode. A voltage pulse is produced by the photodiode that is proportional to the projected surface area of the particle. The resulting voltage pulse amplitude is a nonlinear function of the projected area of the particle, which is the sensed size characteristic. Processing converts this figure to an area-equivalent spherical diameter. Fibers and clear particles (e.g., glass, silica) are not detected. The signals are counted and given a low resolution classification, typically into eight preset approximate size ranges. In this way, the cross-sectional area of the particle is the characteristic size that is measured; the particle size is then specified in terms of the equivalent spherical diameter. The signal is electronically converted to count the number of particles within a given size range, which is preset on the counter. This process is shown graphically in Figure 3-4.

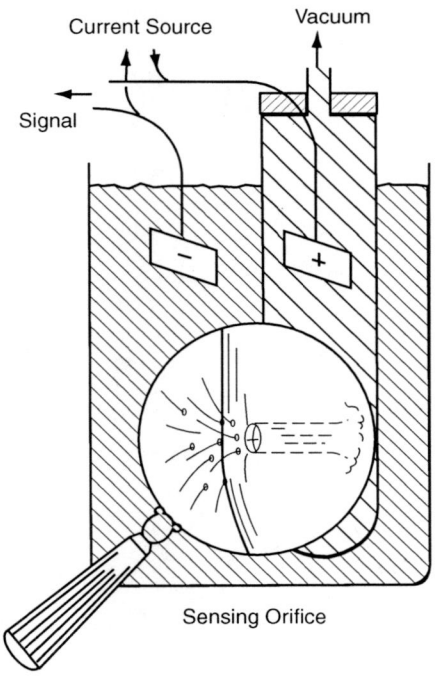

Courtesy of Particle Data, Inc., Elmhurst, Ill.

Figure 3-3   Measuring apparatus of electrical sensing instruments

Equipment can accept a variety of sensors to expand the particle size range that can be evaluated. Individual sensors can detect particle size ranges of about 50:1. The smallest particle size that can consistently be detected with these counters is about 2 μm; below this limit, electronic noise interferes with the instrument's count. For example, the Hiac/Royco sensor model HR-120A has a usable detection range of 2 to 100 μm (Hiac/Royco, 1988). An advantage of this type of counter versus the electrical sensing model is that one orifice is used to measure a wide particle size range. Time savings, therefore, may be obtained by eliminating the need for changing orifices in batch systems.

Laser Light-Scattering Method

The forward-scattering laser light principle has also been incorporated in designs for commercially available particle counters. Hiac/Royco and Spectrex (Redwood City, Calif.) have developed units based on this principle of operation. The two units vary in the way the light-scattering principle is applied. The Hiac/Royco unit draws a sample through a sensor cell through which the laser light is passed. As the particles pass through the light beam, laser light is scattered. The scattered light is passed through a beam deflector and a converging lens, to be picked up by a detector. Based on the principles of Mie-scattering, a calibration curve relates the scattered light intensity to particle size. Hiac/Royco states a detectable size range of 0.4 to 25 μm. This detector can be used in combination with a light-obscuration sensor (although not simultaneously) to extend the combined range of the two sensors from 0.4 to 100 μm.

The Spectrex unit also uses laser light for analysis. It differs from other laser light machines in that the sample remains in its container (usually a glass beaker) during analysis. The rotation of the laser beam provides the necessary screening, and no sample flow is required. The helium–neon laser beam is spatially filtered and focused by two lens assemblies to form an illuminated volume within the liquid sample. A scanning mechanism mounted on the inside of the rear panel provides a lateral displacement of this illuminated volume at a constant speed. The beam is folded by a front surface mirror and two prism assemblies so that it can travel from the rear of the instrument, through the liquid, and to the photodetector assembly at the front of the instrument (Figure 3-5).

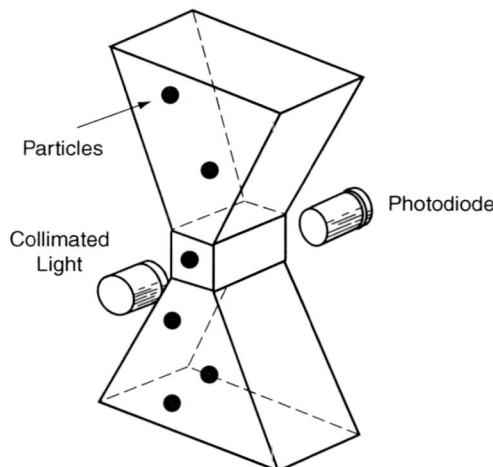

Courtesy of HIAC/ROYCO, Division of Pacific Scientific.

Figure 3-4   Measuring apparatus of light-blockage instruments

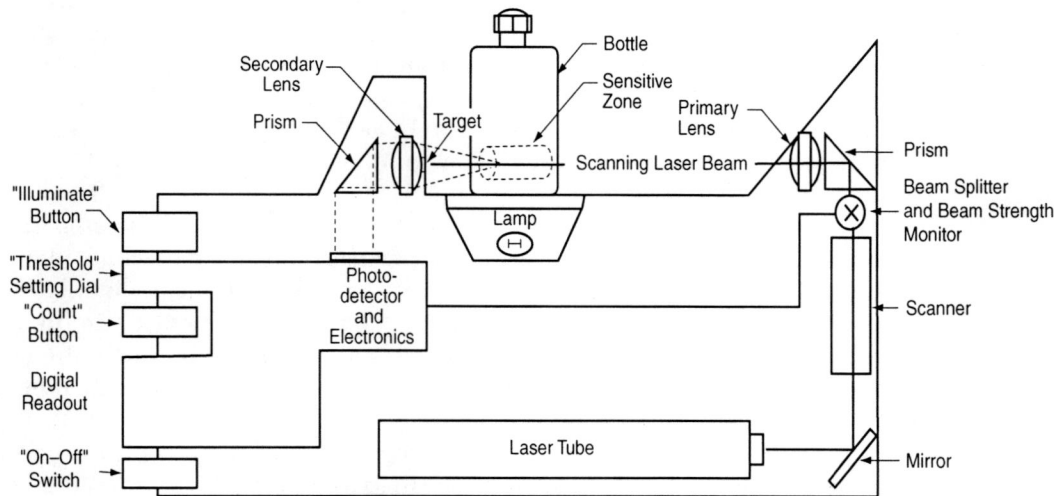

Courtesy of Spectrex Corporation, Redwood City, Calif.

Figure 3-5   Schematic of laser light-scattering particle counter

As the illuminated volume moves across a particle suspended in the liquid, some light from the beam will be scattered. Much of this scattered light is deflected in the near forward direction and is collected by the optical system of the photodetector assembly. The flash of light striking the photodetector causes an electrical pulse in the preamplifier connected to the photodetector. The amplitude of this pulse is a function of the size of the particle. The beam sweeps through the liquid at a constant speed for a set period of time, allowing the instrument to record the number of particles in a predetermined volume. This unit has two lens assemblies, one for analysis in the 0.5 to 16 μm range and one in the 16 to 100 μm range.

METHODS OF OPERATION

Any particle counter is actually a system comprising a sensor, a control unit, and in most cases a computer and software. Currently available particle counters are either on-line or batch instruments.

A batch, or laboratory, instrument requires the collection of a sample in some container, followed by the introduction of the sample to the instrument. Most light-blockage batch instruments use small internal pumps to draw the sample through the instrument for analysis. The electrical resistance instruments require that the sample be treated before analysis. Some models of light-scattering instruments allow analysis in the sample collection container. Data can be recorded manually or stored via computer software.

A batch instrument allows analysis of samples from different locations, and the instrument can be kept in a clean environment. The disadvantages of such an instrument include the potential for sample contamination between the time the sample is collected and analyzed, and the requirement that an analyst must operate the equipment, which limits the number of samples that can be analyzed.

On-line instruments are generally plumbed directly into sample lines, allowing for continuous operation. The samples are continuously analyzed and the data recorded by computer, allowing for more frequent analyses than the batch sampler allows. The main disadvantage of an on-line system is that the system collects samples at only one location. To gather data from more than one filter or more than one location, such as source water and filtered water, an on-line system is required at each location. This configuration increases the overall cost to the utility. Another disadvantage of an on-line system is its limit on the maximum concentration level of particles. For a light-blockage instrument, this limit is generally about 12,000 particles/mL. Most source waters and settled waters exceed that level, so such an instrument cannot be used on-line at such a sampling location. The samples need to be diluted and analyzed with a batch sampler.

Many utilities have purchased batch instruments as a first step in particle counting, in order to evaluate the trends of particles in their particular plants. Other utilities have installed on-line instruments at various locations, including outlets for source, settled, filtered, and finished waters. A batch instrument is useful, even in a plant where on-line instrumentation has been installed, in order to check the calibration of the on-line instruments and to analyze samples with high particle concentrations.

Two AWWA Research Foundation reports deal with optimal operational procedures for both batch and on-line instrumentation in a treatment plant (Hargesheimer et al., 1992, 1995).

OPERATIONAL CONSIDERATIONS

Calibration Variations

Each particle counter is calibrated by factory personnel using latex spheres of known size. Generally, the instruments cannot be calibrated on-site. No standard method of calibration exists for instruments used in water treatment, however, and each manufacturer uses a different procedure. This variation, as well as other differences due to the system construction and operation, results in instruments, even those from the same manufacturer, that don't always match each other when tested side by side on the same water.

This problem has been addressed by the manufacturers, but users must be aware of these problems. Latex spheres are now commercially available (Duke Scientific, Palo Alto, Calif.) that allow utilities to set consistent calibration curves of all instruments used at one plant.

Real-World Complications

Particle counters are calibrated using latex spheres of certified size, but particles in natural water are rarely perfect spheres, and rarely do they have the same refractive index as these spheres. Natural particles, especially biological particulates, tend to clump together, while the calibration spheres do not. When a particle passes through the laser light beam, the electronic pulse produced is converted to an equivalent diameter based on the internal standard of the calibration spheres. For this reason, the particles in natural water are not always sized correctly by particle counters. This problem can be minimized by considering only cumulative size ranges, such as greater than 2 µm, rather than trying to rely on discrete size channels such as 2 to 5 µm, 5 to 7 µm, etc. The following paragraphs detail other real-world complications.

Particle disturbance by shearing. Particle size may be distorted as the particles pass through the orifices and tubing of the instruments. Large, porous aggregates may be subject to "feathering," an effect that distorts and elongates the aggregate as the suspending fluid is passed through the sensing zone at a high flow rate. In this case, one large floc may be counted and sized as several small particles

Distortion of particle distribution by dilution. An upper limit determines the number of particles an instrument can read. If the concentration of particles in a sample passing through a counter's orifice exceeds that limit, then coincident passages may occur in which two smaller particles are counted and sized as one larger one (Kavanaugh et al., 1980). Thus, for medium- to high-turbidity water, samples must be diluted sufficiently to minimize coincident opportunities.

This problem occurs with all types of particle counters, although the concentration limit differs from instrument to instrument. The laser light-scatter instrument (Spectrex) has an upper limit of 1,000 particles/mL, while most of the light-blockage instruments have upper concentration limits of 12,000 to 20,000 particles/mL. If the actual counts exceed those levels, then double counting and shadowing of particles occur.

Counts in most filtered effluents fall below these limits, but those in many samples of source water and settled water will exceed the limits, requiring dilution in order to be analyzed. At present, this adjustment is possible only using a batch instrument; an on-line instrument reads directly from a sample stream, and currently available instruments lack dilution mechanisms. On-line instruments cannot be used in locations with particle concentrations higher than the concentration limits of the instruments.

Although samples can be diluted for analysis by a batch unit, they often must be greatly diluted. For example, using an instrument with an upper concentration limit of 1,000 particles/mL to count a sample containing 50,000 particles/mL, 2 mL of sample would be diluted in 100 mL of particle-free water. Diluting samples for particle counting is problematic because it requires particle-free water, and because large dilution factors often result in inaccurate results.

Particle-free water is very difficult to obtain. Laboratory-grade deionized water is not necessarily free of particles, and filtering water once through a 0.2-µm filter will not produce a diluent of sufficient quality. Most laboratories and utilities that use particle counting develop particle-free water by building closed-loop filtration systems that essentially pass the water through submicron filters many times. Obviously, the need to dilute samples poses an additional problem because of the opportunity to introduce more particles into the sample. Further, work by McTigue and Burman (1988) suggested that highly diluted samples do not produce particle sizes and distributions similar to those of undiluted samples. It is not clear if these distortions are caused by chemical changes in sample makeup, as the diluent, deionized water, becomes the major component of the sample. In situations where extreme dilutions cannot be avoided, it is recommended that the same dilution factor be used for all samples.

Dilutions may also cause distortions in electrical sensing instruments. Samples must be diluted in a strong electrolyte in order to make them sufficiently conductive. In natural freshwater samples, this condition may result in the salting out of certain colloids (Beard and Tanaka, 1977; Lawler, O'Melia, and Tobiason, 1980).

Sample contamination. In general, sample handling for batch analysis is the most difficult aspect of implementing particle counting technology. Any error in handling the sample that introduces contamination will invalidate the readings.

Contamination can be introduced to a sample from the air (since dust particles are approximately the same size as the particulate matter of interest in water), from laboratory glassware, and from dilution procedures and diluent. Samples need to be collected, transported (if necessary), and analyzed in very clean glassware to eliminate false readings from dust and dirt. Glassware must be new, of high grade, and scrupulously clean. It should be cleaned after each use by a combination of autoclaving and ultrasonic means.

Once a sample is collected, care must be taken to avoid any type of contamination. The sample bottle must be handled as little as possible, and no air space may be left. Dust in the air can greatly distort a sample reading, so the sample must be covered from the point of sampling to the particle counter.

Sample holding time.   All samples, particularly source water and flocculated water samples, should be analyzed as quickly as possible after collecting them. Researchers recommend that all samples should be analyzed within 24 h of collection (LeChevallier, 1995).

Sampling time.   To use electronic particle counting as a practical process control tool, the analyses must be performed in a   fairly rapid sequence. For the electrical sensing zone method, a typical time to produce a particle size distribution would be about 30 min to 1 h. The time requirement could be longer in case of extensive plugging of the orifice(s). This type of particle analyzer does not appear to be feasible for on-line measurements due to the need to monitor the orifice for plugging.

Light-obscuration models allow more timely particle size analysis. Generally, several replicate analyses can be performed within 5 to 10 min, unless dilution is required.

Experience with the laser light-scattering model from Spectrex shows that an analysis time of about 5 min is required to complete a particle size distribution. The instrument can switch lens assemblies rapidly, making it very easy to obtain the full range of 0.5 to 100 µm size information.

Cost.   A particle count system is made up of a sensor, a control unit, and usually a computer, computer interface, and software. The price of a system depends on the number of sensors, the equipment configuration, whether it is computer driven, and whether the computer is part of the purchase price. As an example, a manual batch system without computer interface would cost approximately $7,000. A two-sensor on-line system with computer, computer interface, and software would cost approximately $11,000.

Reproducibility.   The precision of various particle counters has been examined extensively in the literature. Thus far, no definite method has been established to correct for all identified problems. However, by understanding the principles of operation and the equipment's limitations, tests can gather information that will enable reasonable interpretation of analytical results. Absolute sizing of many particles in natural waters and at various stages in a treatment plant may not be possible, or even necessary for success. An electronic particle counter provides data to follow the trends of distribution of a broad range of particle sizes through a given treatment process, thus providing valuable information on its effectiveness.

Understanding how a particular instrument works will also help to explain differences among the results found in the literature and in practice. Specific instruments operate on different principles, so data obtained from them may not always be directly comparable.

Differences in data due to type of instrument.   In previous sections, the basic principles used in particle counting instruments were described. Each instrument manufacturer applies these principles in different ways, both in the actual instruments and in supporting software. Because of this variation, each type of particle counter differs slightly from others. As a result, different instruments may not generate directly comparable data.

In October 1988, a study analyzed water samples from the same source and filtered water on six different particle counters (McTigue and Burman, 1988). The samples were taken, split, sent by overnight mail on ice to various locations, and analyzed as soon after receipt as possible. The purpose of the exercise was to determine if the type of instrument and software used affected the reported data. The type of instrument, the reported range of the instrument, and the researchers involved in the study are listed in Table 3-1. The researchers were asked only to analyze the samples in the same manner they would typically use for a raw and a filtered water sample.

Table 3-1   Particle Counters Used in Split-Sample Analysis, October 1988

Instrument	Type	Detection Range	Research and Affiliation
Hiac PC-320	Light obscuration	1–100 µm	J. Cleasby, Iowa State University, Ames, Iowa
Spectrex SPC-410(1)	Laser light scattering	1–100 µm	N. McTigue, Environmental Engineering and Technology, Newport News, Va.
Spectrex SPC-410(2)	Laser light scattering	17–100 µm	K. Burman, Tulsa City Laboratory, Tulsa, Okla.
Particle Data Elzone System	Electrical resistance	0.3–1,200 µm	Particle Data, Inc., Elmhurst, Ill.
Coulter Counter	Electrical resistance	0.5–900 µm	D. Lawler, University of Texas, Austin
Brinkman Particle Size Analyzer	Laser light blockage	0.5–300 µm	A. Amirtharajah, Georgia Institute of Technology, Atlanta, Ga.

Significant differences appeared among the researchers' reported data. Table 3-2 lists the total particle counts in 1 mL, samples of both source and filtered water. The results for the source water sample ranged from 16,000 to 2.3 million particles/mL, and the filtered water results ranged from 378 to 8,300 particles/mL.

The great variation was primarily due to the differences in detection ranges of instruments. The two electrical-resistance machines covered a much broader range than the light-based machines, so it is not usual that those instruments would give much higher counts for the source water. This effect of detection range is also apparent in the different readings the two Spectrex machines obtained. One was set to read from 1 to 100 µm and the other from 17 to 100 µm.

Methods of data reporting.   No standard method applies for reporting particle count and size distribution data. It is important when collecting and comparing data to note the method of data reporting. Figure 3-6 demonstrates how the same particle size distribution data can be plotted in four different ways, resulting in substantially different plots. This graph shows the distribution in terms of number, length, mass, and surface of particles. Figure 3-7 shows another method of reporting particle size distribution data often seen in the literature. The log of the particle diameters is plotted against the log of the particle size distribution function.

Table 3-2   Total Particle Counts of Split Samples, October 1988

Instrument	Source Water Samples number of particles / mL	Filtered Water Samples number of particles / mL
Brinkman	700,000	8,300
Hiac	204,618	1,160
Particle Data	2,282,564	Not analyzed
Spectrex 1	16,000	418
Spectrex 2	140,000	378
Coulter Counter	1,500,000	4,200

Most particle counts are reported as the number of particles in 1 mL of sample. Note, though, that such a value actually means the number of particles in 1 mL detectable by the particular machine used. Particle count and size data require careful interpretation.

Note another difference between Figures 3-6 and 3-7. Figure 3-6 shows each data point as it was collected during a filter run. Figure 3-7 shows results for the same filter run, but the data have been reduced to a probability plot. Reporting the data in this way allows for the run to be fully described by the median, or 50th percentile, as well as the 10th and 90th percentiles. These three data points are easier to analyze than all of the data shown in Figure 3-6 (McTigue, LeChevallier, and Clancy, 1997).

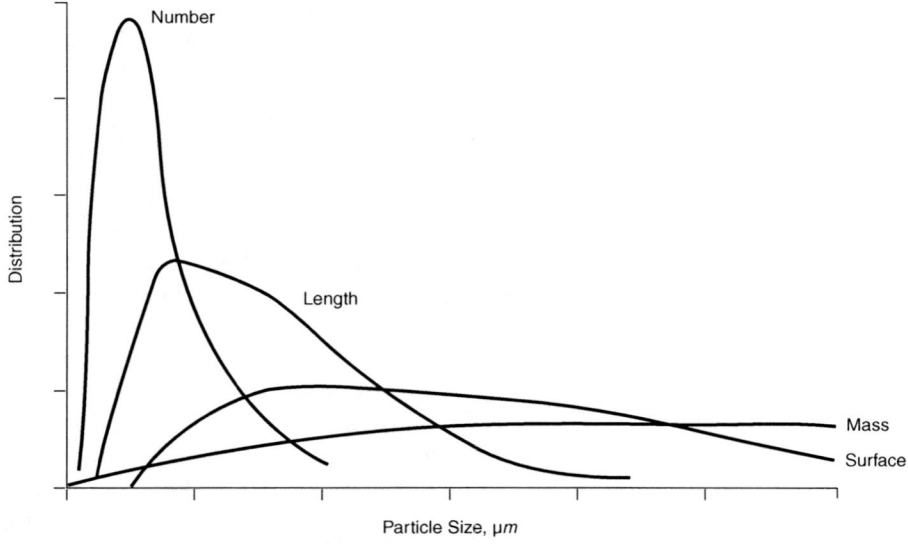

Source: Reprinted with permission from Kavanaugh, M.C. et al. 1980. Use of Particle Size Distribution Measurements for Selection and Control of Solid / Liquid Separation Process. In Particulates in Water: Characterization, Fate, Effects, and Removal, M.C. Kavanaugh and J.O. Leckie, eds. Advances in Chemistry Series 189. Copyright 1980, American Chemical Society.

Figure 3-6   Types of particle distributions

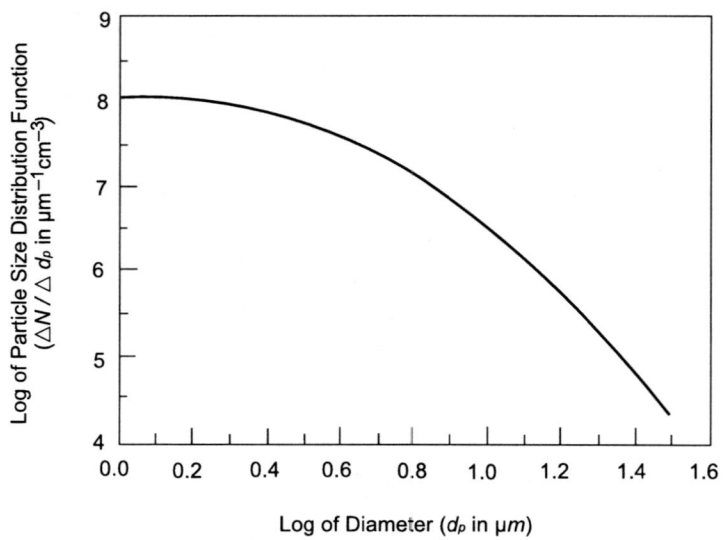

Source: Lawler, Izurieta, and Kao (1983).

Figure 3-7   Hypothetical particle size distribution

Data storage and interpretation.   A particle counter, especially an on-line system, generates a great deal of data. For example, an on-line system can be set to read samples every 10 min. The counter can generate data in a number of size ranges, typically six, and data can be collected at numerous points in a water treatment plant. Collecting this quantity of data every day presents the problems of data storage, interpretation, and handling. Each particle counter manufacturer has its own data storage system; a utility user must be certain that the type of data storage offered is appropriate for the particular use of the particle counters. At a minimum, data from an on-line system should be stored in reports that show trends during each filter run.

OPERATING EXPERIENCE

Optimizing Plant Performance Using Particle Counters

Many published reports have described optimized treatment performance through the use of particle counting. In some cases, particle counters have been used on settled water to determine floc size and distribution. In general, however, the most useful method for optimizing treatment in a full-scale plant is by monitoring the filtered water quality. As discussed earlier in this chapter, organisms of concern to the water industry are actually particulates in the size ranges that particle counters measure. No direct correlation has been made between the concentration of pathogens in raw or filtered water and the number and size of particles. Results from a nationwide survey of 100 water plants suggested that lowering the number of particles in filtered effluent can lower the risk of pathogen presence (McTigue, LeChevallier, Clancy, 1997).

This study also provided information about the factors that affect filtered water particle counts. Using light-blockage on-line particle counters, the median filtered water particle count of particles greater than 2 µm was less than 20 particles/mL at these 100 plants. But 90 percent of the plants experienced periods when filtered water particle counts "spiked," that is, the counts were much higher over specific periods than the median counts of the overall filter runs. Most of these spikes were caused by filter ripening, but many spikes were also observed during the middle and end portions of filter runs. Causes of most of these periods of high particle counts were traced to operational changes in the plants, such as changes in filter loading rates, backwashing of adjacent filters, and variations in chemical feed. Particle counters allow operators to detect these spikes and possibly eliminate them through changes in operations.

Full-Scale Application Case Study

The Alfred Merritt Water Treatment Facility of the Southern Nevada Water System utilizes on-line particle counting for monitoring and control of filtration. Hutchinson (1984) described this system in detail. This section reviews results reported in this paper.

The plant is part of the system that serves the Las Vegas, Nev., area. It has a maximum treatment rate of 600 ft^3/s (17 m^3/s), making it one of the world's largest direct-filtration facilities. It is the first plant in the world to use a continuous on-line particle counting system. This system is used to monitor the effectiveness of the treatment process and to initiate automatic backwashing of filters.

A laboratory particle counter had been used on an experimental basis prior to the expansion of the treatment plant. This equipment proved to offer a good method for plant personnel to optimize plant performance while limiting chemical dosages.

The operator tested the control system on a small pilot filter that was a scaled replica of the treatment process. Testing varied key parameters such as flocculator energy input, chemical mixing times, chemical dosage, and filtration rate. The lab particle counter then permitted the operator to compare the number and size of particles in the pretreated and finished water to establish optimal chemical doses. Figure 3-8 illustrates how an optimal alum dosage could be determined based on a curve fit of several pilot filter runs. The direct result of implementing this procedure was to reduce chemical costs by approximately 32 percent. Additional savings were realized by a reduction of filter backwash frequency and a corresponding load reduction on the residuals-handling system.

This experience led to the decision to add on-line particle counters during the plant expansion to allow further optimization of the full-scale treatment process through improved chemical dosage control and quality monitoring of the treated effluent.

The on-line particle counting system was designed to analyze particle number and size distributions at every key step in the treatment process. The sampling points were stationed in the source water immediately after predisinfection, prior to flocculation, prior to filtration, in the effluent streams of all 20 filters, and on the combined plant effluent flow. The objective of this monitoring system was to enable the operator to develop an optimal filtration particle size that gives deep filter media penetration. This objective is greatly facilitated by computer displays that allow the operator to observe changes in total particle number and size distributions through the various treatment processes. For example, the agglomeration of particles through the flocculation process and final separation in the filter media can readily be monitored and modified by adjustment of coagulant doses, flocculator energy input and detention time, etc. The operator can thus prevent problems such as excessive coagulation, which forms a large, visible floc that remains on the filter media surface as opposed to the microscopic floc needed for maximum bed penetration and maximum filter run times.

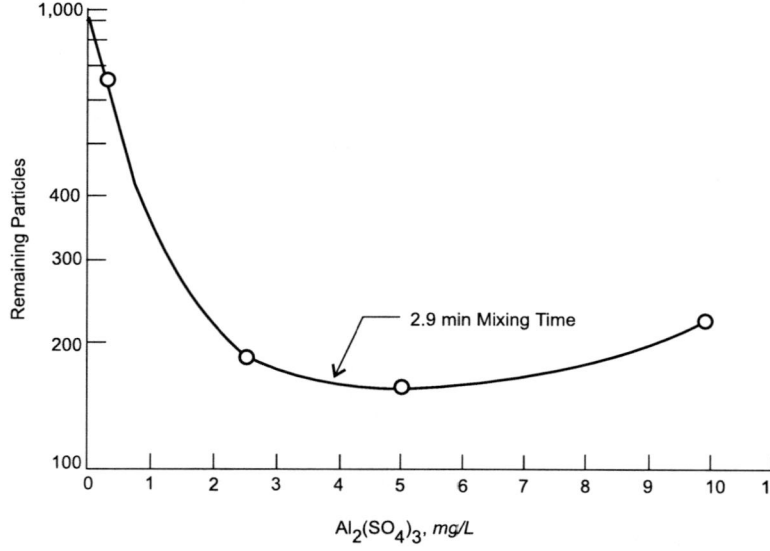

Figure 3-8 Optimal alum dose determined based on pilot runs

A primary treatment objective is to limit the overall particle count to less than 20 particles per millilitre of water, at which time automatic filter backwash is initiated by the computer system. Criteria to trigger backwash were selected to be 150 h of filter run or over 8 ft (2.4 m) of filter head loss.

Single sample streams such as source water and finished water are alternately fed to an on-line volumetric sampler (OLVS). The OLVS takes a volumetric sample and directs it through the sensor at low pressure and flow rate. The standard particle sensor range is 2.5 to 150 μm. However, alternative sensors are available for flocculants up to 600 μm. In other cases with multiple sampling points, a multiple sensor selector (MSS) is provided to enable a single particle counter to receive data from up to 10 OLVS units. Data from each individual sensor are presented in sequence to the particle counter to enable it to complete its tests.

This plant has operated with on-line particle counters since 1984. Operational results have been good, and the utility is considering using these instruments in its next expansion.

SUMMARY

Particle counting and sizing techniques give greater insight into the water treatment process than other current techniques provide. With increasingly stringent regulations governing treated water quality, detailed information on source water quality and the performance of unit processes will gain importance. Particle counters can give the plant operator the information needed to optimize coagulation and filtration. Recognition of the instruments' limitations is critical, however, as discussed in this chapter, to ensure appropriate decisions using the data generated.

REFERENCES

Arora, H., M.W. LeChevallier, and W.D. Norton. 1995. The Relationship Between Particle Counting and Parasites: Full-Scale Experience. *Proc. AWWA Annual Conference.* Anaheim, Calif. Denver, Colo.: American Water Works Association.

Beard, J.B., III, and T.S. Tanaka. 1977. A Comparison of Particle Counting and Nephelometry. *Jour. Awwa,* 69:10:533

Berman, D., E. Rice, and J. Hoff. 1988. Inactivation of Particle-Associated Coliforms by Chlorine and Monochlorine. *Appl. & Envir. Microbiology,* 54:2:507.

Hargesheimer, E.E., C.M. Lewis, and C.M. Yentsch. 1992. *Evaluation of Particle Counting as a Measure of Treatment Plant Performance.* Denver, Colo.: American Water Works Association and AWWA Research Foundation.

Hargesheimer, E.E., and C.M. Lewis. 1995. *A Practical Guide to On-Line Particle Counting.* Denver, Colo.: American Water Works Association and AWWA Research Foundation.

Hiac/Royco. 1988. *Contamination Monitoring-Sensors to Size and Count Particles.* Silver Spring, Md.: Hiac/Royco Instruments Div.

Hutchinson, C.W. 1984. On-Line Particle Counting Improves Filter Efficiency. Proc. ISA Intl. Conf. and Exhibit. Houston, Texas: Instrument Society of America.

Kavanaugh, M.C., et al. 1980. Use of Particle Size Distribution Measurements for Selection and Control of Solid/Liquid Separation Process. In *Particulates in Water: Characterization, Fate, Effects, and Removal.* M.C. Kavanaugh and J.O. Leckie, eds. Advances in Chemistry Series 189. Washington, D.C.: American Chemical Society.

Lawler, D., E. Izurieta, and C.P. Kao. 1983. Changes in Particle Size Distributions in Batch Flocculation. *Jour. AWWA,* 75:12:604.

Lawler, D., C. O'Melia, and J. Tobiason. 1980. Integral Water Treatment Plant Design: From Particle Size to Plant Performance. In *Particulates in Water.*

M.C. Kavanaugh and J.O. Leckie, eds. Advances in Chemistry Series 189. Washington, D.C.: American Chemical Society.

McTigue, N.E., and K. Burman. 1988. Particle Counting for Water Treatment Plant Control. *Proc. AWWA Water Quality Technology Conf.* St. Louis, Mo. Denver, Colo.: American Water Works Association.

McTigue, N.E., M. LeChevallier, and J. Clancy. 1997. *National Assessment of Particle Removal by Filtration.* Denver, Colo.: American Water Works Association and AWWA Research Foundation.

Saunier, B.M., R. Selleck, and R.R. Trussell. 1983. Preozonation as a Coagulant Aid in Drinking Water Treatment. *Jour. AWWA,* 75:5:239.

Tate, C.H., and R.R. Trussell. 1978. The Use of Particle Counting in Developing Plant Design Criteria. *Jour. AWWA,* 70:12:691.

Chapter **4**

Electrophoretic Mobility Measurements

The engineering literature contains numerous references to the use of electrophoretic mobility (EM) or zeta potential (ZP) measurements in coagulation process control techniques. Many of these papers describe attempts to correlate ranges of EM values or changes in the EM, such as sign reversal, with the efficiency of particle removal by flocculation followed by sedimentation and filtration. This chapter describes how the EM measurement is made and discusses how and when it can be used to control and facilitate coagulation process operation.

BACKGROUND

Electrophoresis is an electrokinetic effect, so it is explained by the same fundamental principles that account for other electrokinetic phenomena, such as streaming current (or streaming potential), sedimentation potential, and electro-osmosis. In electrophoresis, particles suspended in a liquid are induced to move by the application of an electric field across the system. This technique has been used by colloid chemists for many years to determine the net electric charge or near surface (zeta) potential of particles with respect to the bulk of the suspending phase.

EM measurements have long been used in methods to control applications of coagulant chemicals in solids–liquid separation systems. Prominent water treatment researchers have published a number of papers on this topic. Despite support from convincing advocates, however, inconsistent and difficult-to-interpret results and the time-consuming nature of EM determination appear to have limited the widespread use of EM measurements in routine treatment plant operations.

MEASUREMENT TECHNIQUES

Microelectrophoresis

A number of different techniques are used to determine particle EM. The most important of these in water treatment applications has been microelectrophoresis. In this method, the suspension is contained in a small-diameter glass or plastic tube that has, in most cases, a round or rectangular cross section. An electric field is applied across the contents of the tube in the axial direction using a stable, constant-voltage power supply and inert (e.g., platinized platinum) electrodes inserted in sealed fluid reservoirs at the ends of the tube. When the voltage is applied, the particles tend to migrate in the axial direction. The EM is determined by measuring the average velocity of particle migration and dividing this value by the voltage gradient across the electrophoresis cell. The voltage gradient is determined by dividing the applied voltage by the effective length of the cell.

Velocity Measurement

A number of methods are used to measure the average velocity of particle migration in an electrophoresis cell. The oldest technique involves the use of a light microscope with dark field illumination to enhance particle image resolution. The microscope is used to observe the motion of single particles, and the velocity is determined by timing their passage between lines on a calibrated eyepiece graticule. During measurements of the velocities of particles, the polarity of the electrodes is reversed between measurements to minimize their polarization. Polarization tends to cause an uncontrolled variation in the voltage gradient across the cell. The light microscope method is employed in the instruments sold by Zeta Meter Corp. (New York) and Rank Bros. (Bottisham, Cambridgeshire, England).

More elaborate methods are available for measuring particle velocity, but they invariably require more expensive equipment than light microscope systems, while allowing for more rapid and accurate analysis. The prices of light microscope systems start at about $8,000; the higher-priced instruments range from $10,000 to $20,000. An apparatus sold by the Pen Kem Co. (Croton-on-Hudson, N.Y.) incorporates a mechanism in which the operator varies the rotational speed of a prism. Light scattered by the moving particles passes through the prism; when the rotation of the prism just compensates for the motion of the particles, they appear to be stationary. The prism control knob is calibrated to give a direct reading of EM. Particle illumination in this system is provided by a laser.

More elaborate devices are available for EM measurement; however, most are designed for research applications. The Pen Kem Co. sells an automatic instrument that uses a multistage photomultiplier tube and a frequency tracker circuit to determine average particle velocity. Several newer instruments use laser light illumination of the particles and measure the electrophoretic velocity by detecting the Doppler shift of the scattered light. The laser Doppler approach has been found to be compatible with cell designs and operating methods that help minimize complicating factors such as Joule heating of the cell contents and fluid motion due to electro-osmosis (Ware, 1977).

Some instruments couple video cameras to microscopes focused on the test cells. The video image of the moving particles is fed to a computer that applies dynamic image analysis to determine the velocity of each particle in the field of view. The Zeta Reader apparatus (sold at one time by Komline-Sanderson Inc. of Peapack, N.J., for use in water treatment plants) uses a manual process in place of the computer. The operator adjusts a dial until a vertical line on the video screen moves across the screen as fast as a selected particle. The dial setting gives the EM.

Electro-Osmosis

In a conventional electrophoresis cell, measurement of the electrophoretic velocity is often complicated by fluid motion caused by the charge on the surface of the cell. If the surface of the cell is negatively charged, the positive countercharge in the fluid next to the surface causes an overall movement toward the negative electrode when the electric field is applied. If, as is usually the case, the cell is sealed at both ends, movement of fluid in one direction produces a return flow in the opposite direction. Consequently, depending on a particle's position in the cell, the measured velocity may not be attributable entirely to electrophoresis.

The most widely used technique for minimizing the effect of electro-osmotic flow involves focusing the microscope (or the laser beam in some laser Doppler instruments) on a point in the electrophoresis cell where the fluid velocity is zero. For simple cell geometry, these points, called *stationary points*, are a distance from the cell walls that can be determined using principles of hydrodynamics. Stationary point locations are usually identified in instrument documentation.

It is possible to reduce or eliminate electro-osmotic flow by coating the walls of the cell with a substance (usually a polymer) that minimizes the surface charge. While this procedure can be used in certain biological systems, its application is not likely to be of value when inorganic colloids are present. If the coating material adsorbs on the glass of the cell, it is likely to adsorb on inorganic particles, as well.

Effective Length of the Cell

The effective length of an electrophoresis cell, as noted earlier, is used with the applied voltage to determine the voltage gradient across the cell. To determine the effective length, the cell is filled with a solution of known electrical conductivity C, and its resistance R is determined with an AC resistance bridge. The effective length l is given by the formula:

$$l = R \times C \times A \tag{Eq 4-1}$$

A is the cross-sectional area of the cell at the point where the velocity measurement is made. This value is usually determined by direct microscopic measurement of the inside dimensions of the cell.

Zeta Potential Determination

The relationship between the ZP of the particle–solution interface and the measured EM may be simple or complex, depending on the magnitude of the quantity Ka, where K is the Debye–Hückel parameter and a is the radius of the particle. The magnitude of K is given by the formula:

$$K = [e^2/(\varepsilon kT)]^{\frac{1}{2}} \; [I]^{\frac{1}{2}} \tag{Eq 4-2}$$

Where:

e	=	the charge on the electron
ε	=	the permittivity of the suspending liquid
k	=	Boltzmann's constant
T	=	the absolute temperature
I	=	the ionic strength of the solution

For 25°C and particles suspended in water:

$$K \, (nm^{-1}) \; = \; 3.26 \; I^{\frac{1}{2}} \qquad\qquad \text{(Eq 4-3)}$$

For values of Ka greater than approximately 200 (i.e., situations with large particles and/or high ionic strength), ZP is given by:

$$ZP \; = \; (\mu/\varepsilon) \; EM \qquad\qquad \text{(Eq 4-4)}$$

Where:

μ =   the absolute viscosity

The relationship for values of Ka less than about 0.1 is

$$ZP \; = \; (3/2) \; (\mu/\varepsilon) \; EM \qquad\qquad \text{(Eq 4-5)}$$

For intermediate values of Ka ($0.1 < Ka < 200$), the relationship between ZP and EM is complex and depends on Ka, the magnitude of ZP, and the mobility levels of the electrolyte ions. When ZP is less than approximately 20 mV, the Henry equation can be used, with $f(Ka)$ as a correction factor:

$$ZP \; = \; [(3/2)(\mu/\varepsilon)](EM)/[1 + f(Ka)] \qquad\qquad \text{(Eq 4-6)}$$

Some values of Ka, $f(Ka)$ are 0, 0; 0.1, 0.001; 1.0, 0.027; 5, 0.160; 10, 0.239; 50, 0.424; 100, 0.458, and ∞, 0.5. A detailed analysis of the relationship between ZP and EM is given in Hunter (1981).

In microelectrophoresis measurements where the particle velocity is determined using light microscopy, the particles are usually larger than 1 μm (1,000 nm) and the ionic strength is 0.001 or greater. According to Eq 4-3, these conditions correspond approximately to $Ka = 30$. Using Eq 4-6 for EM in microns per second per volt per centimetre and a temperature of 25°C, the ZP in millivolts will range from $14.8 \times EM$ for $Ka = 32$ to $12.8 \times EM$ for $Ka = 200$.

ELECTRICAL DOUBLE LAYER

The charges on the surfaces of most naturally occurring particles (organic and inorganic) are caused by the ionization of surface functional groups. At low pH, the surface sites tend to be protonated and positively charged; at high pH, ionization of the sites tends to give the surfaces a net negative charge. Most naturally occurring particles are negatively charged in the pH range 5 to 10.

Figure 4-1 is a schematic diagram showing the distribution of charge and potential near the surface of a negatively charged particle. The negative charge just within the surface of the particle is balanced by an excess of positive charge in the solution next to the surface. The location of the positive countercharge is determined by the affinity of these ions for the negative surface. The ions are drawn toward the surface by electrostatic attraction and, in some cases, by a tendency to form weak bonds with surface sites. A countercharge that is held close to the surface by weak bonds is called the *Stern* or *fixed-layer charge*.

The counterions outside the Stern layer are part of what is called the *diffuse layer*. In the diffuse layer surrounding a spherical particle, the distribution of charge in the radial direction is determined by a balancing of electrostatic attraction and the tendency for the ions to move away from the surface by diffusion. Diffusion has essentially no direct effect on the ions in the Stern layer.

The concentration of diffuse-layer countercharge is greatest next to the Stern layer. Within a distance (e.g., 10 nm) that is related to the Debye–Hückel parameter K and the density of charge on the surface, the concentration of countercharge decreases to essentially that of the bulk solution.

It is not known exactly how the measured ZP fits within this picture of the electrical double layer. It is generally assumed that ZP is determined by the charge contained within a sphere that includes a layer of solution in addition to the particle. While it is reasonably certain that the zeta potential does not equal the surface potential of the particle (see Figure 4-1), it is not clear at just what distance from the particle surface the ZP pertains.

A reasonable and very useful assumption is that the ZP is equal to the diffuse-layer potential, Ψ_d, at the point where the diffuse layer meets the Stern layer. This assumption facilitates the use of ZP measurements in the calibration of surface chemical equilibrium models (models that include the diffuse-layer potential as a parameter). It also helps in understanding the relationship between ZP and the tendency of particle suspensions to flocculate or deposit on stationary surfaces.

SIGNIFICANCE OF ZETA POTENTIAL IN PARTICLE AGGREGATION AND DEPOSITION

The tendency of suspended particles to aggregate and attach to a stationary surface (that is, to deposit) is related to the ZP values of the interacting surfaces. Unfortunately, this relationship is not always simple, and the application of ZP measurements for coagulant dosage control may not be a straightforward process.

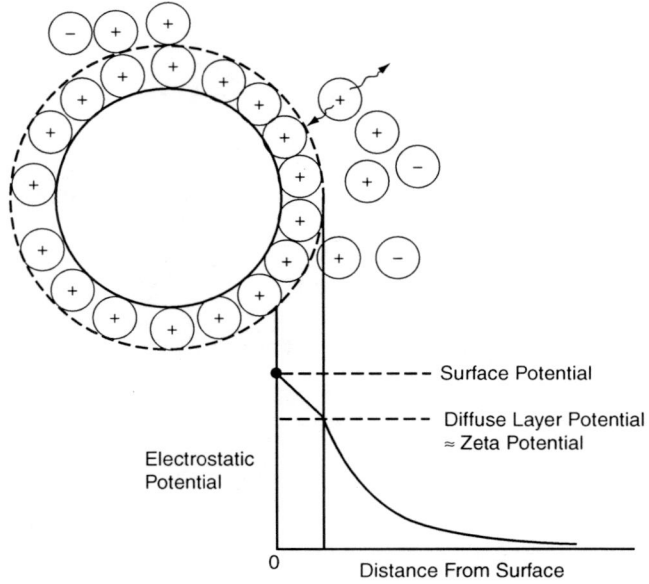

Figure 4-1   Schematic diagram showing the distribution of charge and potential near a negatively charged particle

Repulsion

When two similar particles approach on a line connecting their centers, a number of forces determine whether or not the kinetic energy of the particles will bring them close enough together for contact to occur. As the diffuse-layer ions (see Figure 4-1) intermix, a repulsive force (called *double-layer repulsion*) develops that tends to keep the particles apart. The significance of double-layer repulsion at any separation distance is determined by the diffuse-layer potential, the ionic strength, and the radius of the particle. For a given particle size and separation distance, the repulsive potential energy increases with increasing diffuse-layer potential and decreasing ionic strength.

A second repulsive force called *hydrodynamic retardation* is caused by the viscous displacement of fluid from between the particles. Hydrodynamic retardation is especially important for comparatively large particles.

Attraction

The two repulsive forces are at least partly offset by a third force caused by the London–van der Waals interaction between the approaching particles. The London–van der Waals attractive force tends to pull the particles together as the distance between them becomes very small. This attraction is electromagnetic in character and is caused by the interaction of permanent or electrically induced dipoles in the interacting particles.

Total Interaction Curve

Combining double-layer repulsion and London–van der Waals attraction produces a total interaction potential energy curve (Figure 4-2). The repulsive potential energy V_R decreases exponentially with separation distance, while the attractive potential energy V_A has an approximately inverse relationship with the square of the distance. When double-layer repulsion is significant, the total potential energy curve $(V_R + V_A)$ tends to pass through a maximum. This maximum is the potential barrier that the interacting particles must surmount if lasting contact is to occur. If the kinetic energy

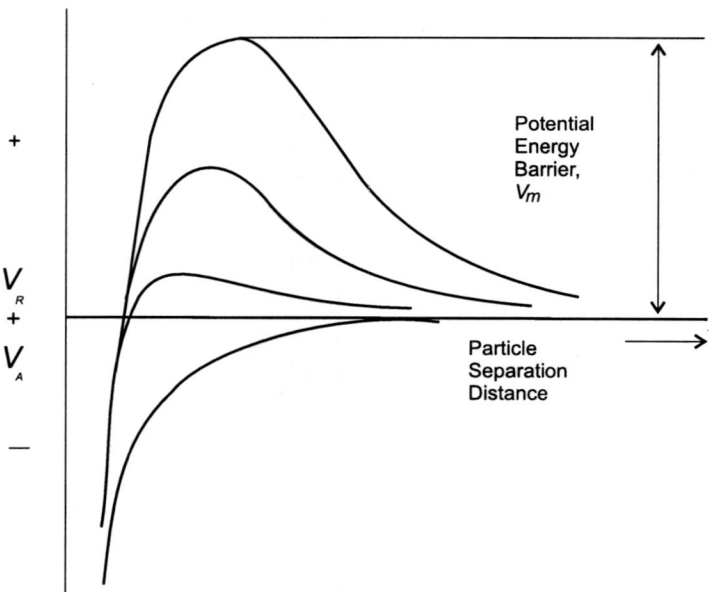

Figure 4-2   Total potential energy curve for the interaction of identical charged particles

of the particles is less than the barrier strength, relatively few particles will make contact, producing a stable suspension.

The magnitude of the potential energy barrier is determined to a large extent by the VR term, so it is a function of the diffuse-layer potential. It is this relationship that establishes a theoretical basis for linking the ZP (which, as noted above, is believed to be at least approximately equal to the diffuse-layer potential) to suspension stability. In general, it can be assumed that as ZP decreases, double-layer repulsion and the maximum in the total potential energy curve also decrease, and the tendency for aggregation and deposition increases. To answer the operationally important question about what values of the ZP are needed to yield rapid aggregation and deposition, an examination of the transport processes that tend to bring the surfaces together is required.

TYPES OF FLOCCULATION

The purpose of flocculation is to promote the interaction of small particles and to form aggregates that can be efficiently removed in subsequent separation processes. Two types of flocculation—Brownian and orthokinetic—will be discussed here.

Brownian Flocculation

Brownian movement of a colloidal particle is caused by the thermal energy of the suspending liquid. The rate of aggregation of two particles (1 and 2) by Brownian movement is given by the expression:

$$I_{12} = 4\pi\alpha_{12}D_{12}(a_1 + a_2)\,n_1 n_2 \qquad \text{(Eq 4-7)}$$

Where:

α_{12}	=	the collision efficiency factor
D_{12}	=	the coefficient of relative diffusion
a_1 and a_2	=	the particle radii
n_1 and n_2	=	the particle number concentrations

The magnitude of α_{12} is a function of the height of the potential energy barrier Vmax (see Figure 4-2), the ionic strength, and the particle radii. This magnitude varies between approximately 0 and a number slightly greater than 1.

The collision efficiency factor has been plotted versus the particle diffuse-layer potential in Figure 4-3 for a particle radius of 45 nm, a temperature of 25°C, and three values of the ionic strength. The diffuse-layer potential, as noted above, determines the height of the potential energy barrier, and therefore the collision efficiency factor. In this example, as the diffuse-layer potential increases, a12 decreases from a value of about 0.93. The curves in Figure 4-3 were plotted using relationships derived by Kim and Rajagopolan (1982).

The purpose of the example in Figure 4-3 is to show that Brownian flocculation displays no single critical value of ZP (the diffuse-layer potential) when the ionic strength of the solution varies. According to the curves plotted in Figure 4-3, as ionic strength increases, the diffuse-layer potential at the point where the collision efficiency begins to decrease (where the suspension just begins to become stable)

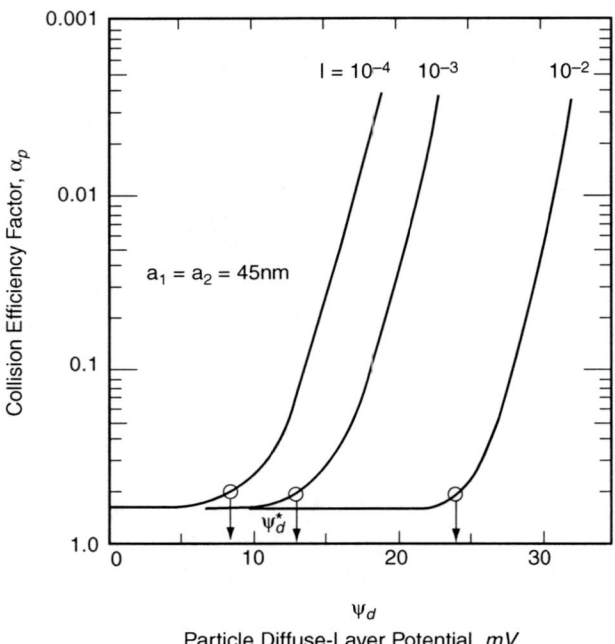

Figure 4-3 Relationship between the diffuse-layer potential and the Brownian collision efficiency factor for a particle radius of 45 nm, a temperature of 25°C, and three values of ionic strength (based on theoretical calculations using equations by Kim and Rajagopolan, 1981)

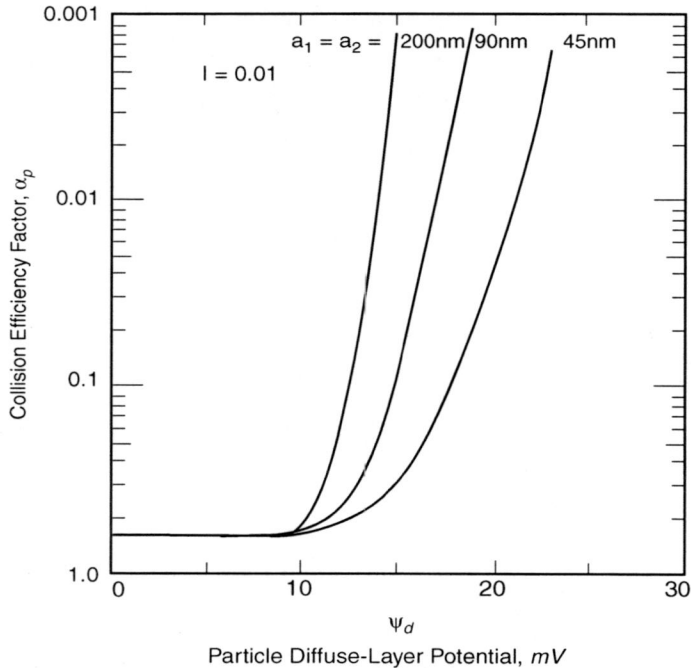

Figure 4-4 Effect of particle size on the relationship between the Brownian collision efficiency and diffuse-layer potential

increases, from about 8 mV for $I=0.0001$ to 24 mV for $I=0.01$. An empirical equation known as the Eilers and Korff Rule describes the relationship between the critical diffuse-layer potential Ψ^*_d and the ionic strength:

$$\Psi^*_d / I^{1/2} = \text{Constant} \qquad \text{(Eq 4-8)}$$

The effect of particle size on the relationship between α_{12} and the diffuse-layer potential is shown in Figure 4-4. For an ionic strength of 0.01, an increase in the particle radius from 45 to 200 nm corresponds to a decrease in the critical diffuse-layer potential from about 15 to 10 mV.

The equations derived by Kim and Rajagopolan (1982) can be used to predict the effect of the diffuse-layer potential on the tendency for submicron-size particles to deposit on a filter surface. Figure 4-5 describes the effect of the diffuse-layer potential on the retention of polio viruses by a microporous membrane filter of 0.45-μm pore size. As the diffuse-layer potential rises from 10 to 30 mV, the fraction of the viruses that pass through the filter increases from 0.001 to approximately 1. It can be shown that the effect of ionic strength and particle size on the critical diffuse-layer potential in this system is similar to that shown in Figures 4-3 and 4-4 for Brownian flocculation.

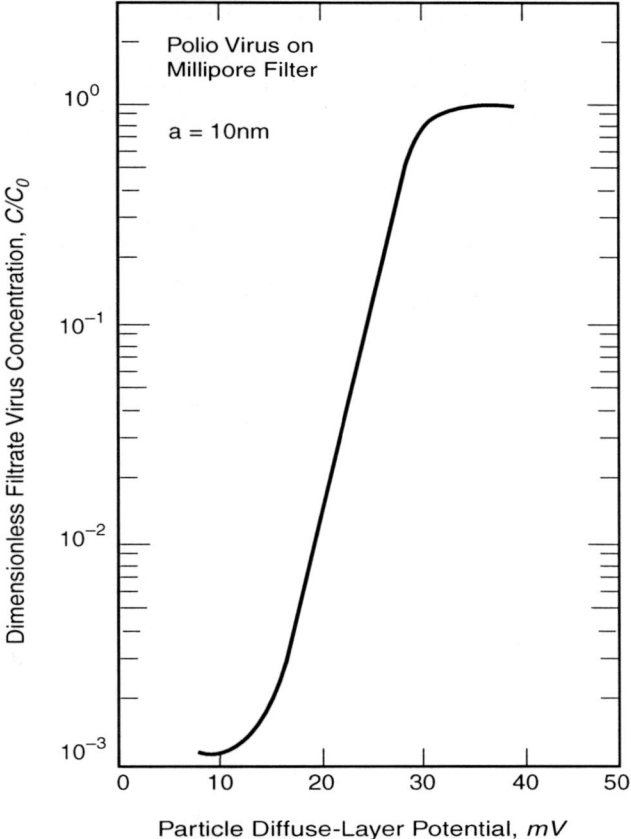

Figure 4-5   Effect of the diffuse-layer potential of virus particles on their removal in a microporous membrane filter

Orthokinetic Flocculation

Orthokinetic flocculation is caused by the relative motion of particles entrained in uniform shear flow. The rate of collisions between equal-size particles is given by

$$I = (8/\pi)\alpha_o \varphi G n$$

(Eq 4-9)

Where:

α_o = the so-called *collision efficiency factor*

φ = the volume concentration of the suspension

G = magnitude of the velocity gradient

n = the particle number concentration

The forces that affect particle interaction in Brownian flocculation (double-layer repulsion, hydrodynamic retardation, and van der Waals attraction) are also important in orthokinetic flocculation. However, in orthokinetic flocculation the force that tends to move the particles together is due primarily to the relative velocity caused by the shear flow and not, as in Brownian transport, by the thermal energy of the fluid.

Current understanding of flocculation in simple shear has been increased by studies employing computer simulations of particle–particle interactions, called *trajectory analysis* (van de Ven and Mason, 1977; Zeichner and Schowalter, 1977). Results obtained by van de Ven and Mason (1977) can be used to determine the effect of the diffuse-layer potential on the magnitude of α_o. Figure 4-6 is a plot of α_o versus the square root of the ratio of two dimensionless parameters, C_R and C_A, which are the ratios of repulsive and attractive forces, respectively, to hydrodynamic forces. The quantity $(C_R/C_A)^{1/2}$, plotted on the x-axis of Figure 4-6, is proportional to the diffuse-layer potential.

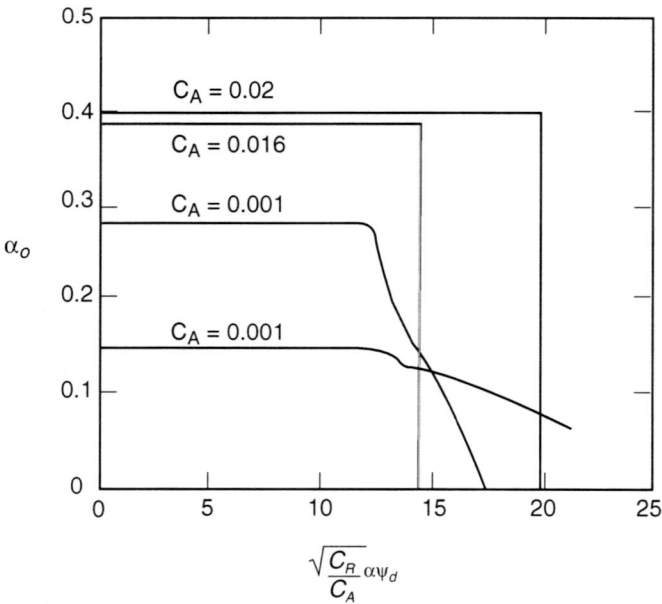

Figure 4-6   Effect of the diffuse-layer potential on the orthokinetic flocculation (simple shear) collision efficiency for several values of van de Ven and Mason's (1977) dimensionless attraction parameter

According to Figure 4-6, when C_A is greater than about 0.001, α_o decreases abruptly from 0.4 to 0 while $(C_R/C_A)^{1/2}$ increases to a value slightly greater than 14. This situation corresponds to the point where the suspension will become stable with increasing diffuse-layer potential (increasing double-layer repulsion). When C_A is less than about 0.001, i.e., when the attractive force becomes less significant compared to the hydrodynamic forces caused by the velocity gradient, α_o begins to decrease at a lower point and the transition is less abrupt. These features shown in Figure 4-6 suggest that for orthokinetic flocculation, the critical diffuse-layer potential will decrease as the magnitude of the velocity gradient increases. At high velocity gradients (very low values of C_A), it may be difficult to detect a critical diffuse-layer potential.

POLYELECTROLYTE COAGULANTS

Flocculation and Sedimentation

Addition of a coagulant may destabilize a suspension by one or more of several mechanisms. When the coagulant is a cationic polymer of low to moderate molecular weight such as polydiallyldimethyl ammonium chloride (polyDADMAC) or epichlorohydrin dimethylamine (epiDMA) polyamine, and the particles are negatively charged, destabilization takes place by a process known as *charge neutralization*. In charge neutralization, the coagulant molecules, with associated positive charges, adsorb on particle surfaces and decrease their net surface charges. This activity reduces the diffuse-layer potential and decreases double-layer repulsion. In combination with particle transport (via Brownian diffusion, laminar velocity gradients, or turbulent flow), the effect produces aggregation or deposition. In Figure 4-7, EM and residual turbidity (measured after flocculation and sedimentation in a jar test apparatus) are plotted versus the concentration of polyDADMAC product added to a negatively charged silica suspension. The residual turbidity reaches a minimum at a product concentration of approximately 10^{-4} g/L; this is approximately the dosage at which EM approaches zero. The addition of polyelectrolyte beyond this point causes the EM (and the net charge) of the particles to become positive and the suspension restabilizes.

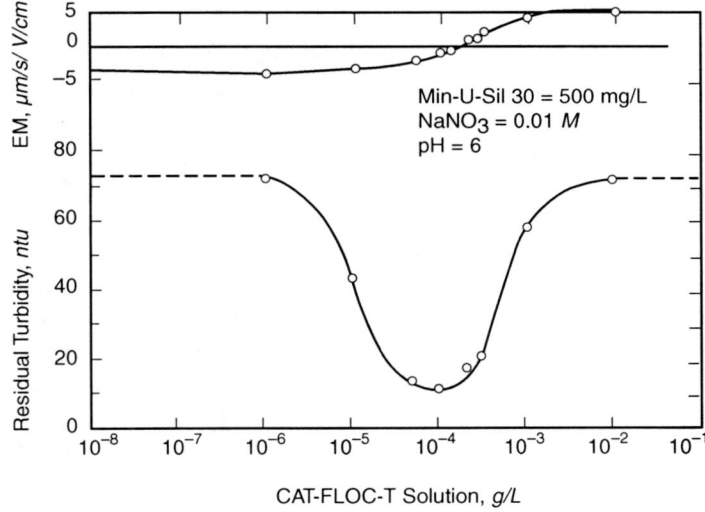

Figure 4-7   Jar test data illustrating the effect of polyDADMAC dosage on electrophoretic mobility and residual turbidity after flocculation and sedimentation

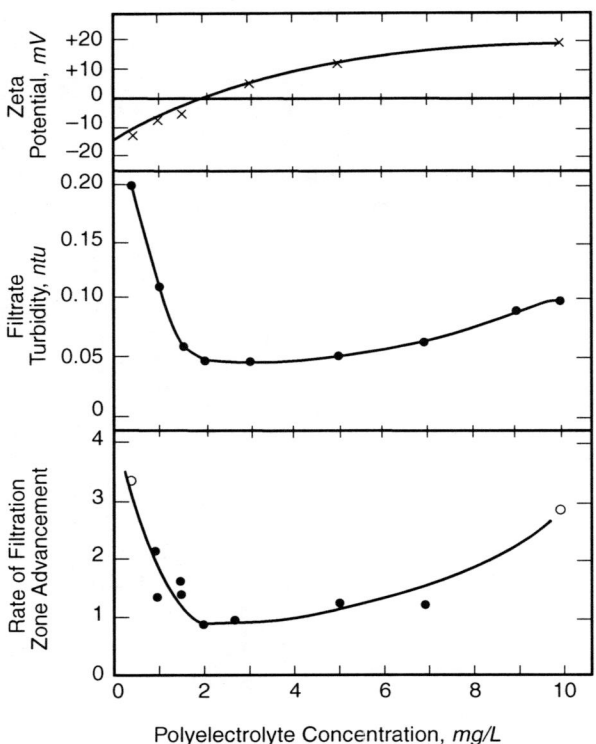

Figure 4-8 Pilot-plant data from a direct filtration system showing the effect of polyDADMAC dosage on electrophoretic mobility, filtrate turbidity, and rate of filtration zone movement into the filter bed

High molecular weight anionic and nonionic polymers are used to alter the strength and sizes of particle aggregates, usually by a process known as *bridging*. Bridging may occur if the polymer molecules are long enough to attach to two or more particles at the same time. EM measurements are of limited value for controlling doses of these compounds, because the effect of the polymers on the electrical double layer is not a controlling factor.

Flocculation and Granular Bed Filtration

Figure 4-8 shows the significance of EM measurements in controlling the granular bed filtration process when a cationic polyelectrolyte coagulant is used in a direct filtration system. According to Figure 4-8, as the polyelectrolyte concentration increases and EM decreases toward zero, the filtrate turbidity at a given volume of water filtered decreases toward a minimum value. The rate at which the filtration zone advances into the filter bed also decreases toward a minimum.

In Figure 4-8, the effect of polyelectrolyte addition on filter performance is determined by two factors: particle aggregation and the efficiency of particle attachment to the filter grains. As the polyelectrolyte concentration increases toward 2 mg/L and EM approximates zero, increases are observed in both particle aggregation occurring after polymer addition but before filtration and the efficiency of particle attachment to the filter grains. Both of these factors tend to increase particle removal and decrease the rate of filtration zone advancement.

At polyelectrolyte concentrations greater than about 2 mg/L, the EM becomes positive, and particle aggregation before filtration declines. However, because the surface of the clean filter grains is negatively charged, the dispersed, positively charged, polyelectrolyte-coated particles deposit efficiently on the filter grains, maintaining a low effluent particle concentration and a slow rate of filtration zone advancement. At polyelectrolyte concentrations approaching 10 mg/L, the excess polymer in the system adsorbs on the filter grain surfaces, and interaction between the positive grains and the positive particles is inhibited by increased double-layer repulsion.

HYDROLYZING METAL COAGULANTS

Salts containing hydrolyzing metal ions, such as Fe^{+3} and Al^{+3}, are widely used as coagulants in water treatment. When these compounds are added to a suspension, the metal ions hydrolyze and adsorb (deposit as particles of precipitate) on the particulate surfaces. The action of the metal hydrolysis products can be similar to that of cationic polyelectrolytes, with destabilization and restabilization caused simply by the adsorption of positive charge on the negatively charged particle surfaces (see Figure 4-7). However, other factors will likely affect the destabilization and flocculation processes. Such factors include physical changes in the suspension due to the formation of the voluminous metal hydroxide precipitate and adsorption of anions such as sulfate on the precipitate coating. Consequently, treatment using hydrolyzing metal coagulants may not allow application of EM measurements for coagulant dosage control.

Charge Neutralization by Hydrolysis Products

The data plotted in Figure 4-9 were obtained under conditions where the action of the aluminum hydrolysis products resembles that of cationic polyelectrolytes (see Figure 4-7). The data are from jar test experiments in which $Al(NO_3)_3$ was used at constant pH (pH = 6) with no added sulfate. Under these conditions, destabilization and restabilization were caused by the adsorption of positively charged aluminum hydrolysis products on the negatively charged silica particles. The presence of the voluminous aluminum hydroxide precipitate at the higher aluminum concentrations had a negligible effect on turbidity removal, because the high positive charge of the precipitate coating minimized particle aggregation.

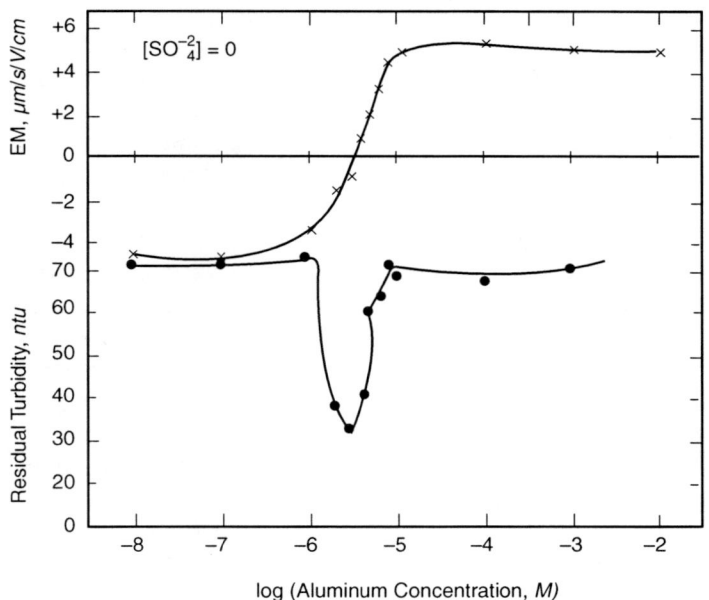

Figure 4-9 Jar test data showing the effect of the aluminum nitrate dosage on electrophoretic mobility and residual turbidity after flocculation and sedimentation for pH = 6 and no sulfate

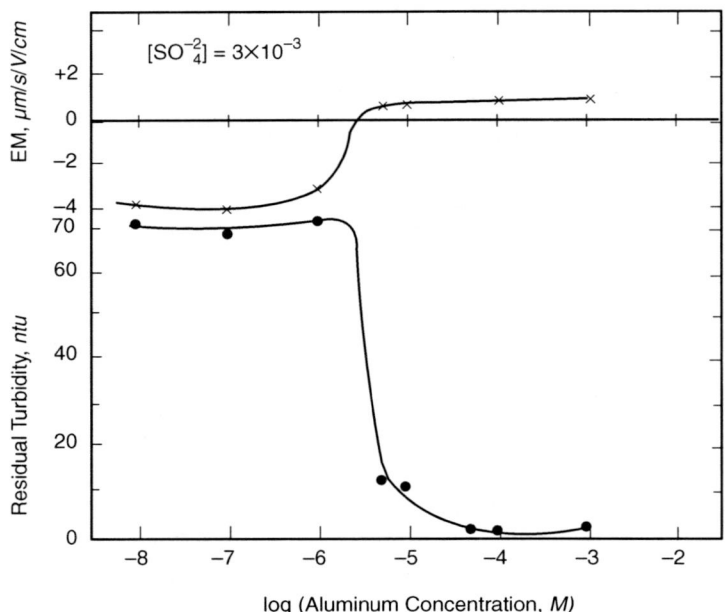

Figure 4-10   Jar test data showing the effect of the aluminum nitrate dosage on electrophoretic mobility and residual turbidity after flocculation and sedimentation for pH = 6 and $[SO_4^{-2}] = 3 \times 10^{-3}$

The experimental conditions for the jar test data plotted in Figure 4-10 were the same as those in Figure 4-9, except that the solution contained 3×10^{-3} M sulfate. The positive charge of the aluminum hydroxide–coated particles was reduced by the adsorption of sulfate ions; with no sulfate, the maximum EM is approximately +5 (Figure 4-9); with 3×10^{-3} M sulfate, the maximum is about +1 (Figure 4-10).

Significance of Enmeshment in the Metal Hydroxide

According to Eq 4-9 for orthokinetic flocculation, the rate of particle interaction is determined by the product of the collision efficiency factor, α_o, and the volume concentration of the suspension, φ. As adsorption of sulfate reduces the surface charge of the aluminum hydroxide–coated particles, this reduction eliminates restabilization at high aluminum concentrations; unlike the no-sulfate case (Figure 4-9), the magnitude of α_o remains high as the aluminum concentration increases. Also, the product $\alpha_o \varphi$ and the efficiency of flocculation increase due to the effect of the aluminum concentration on φ. This effect helps explain why, in the presence of sulfate at aluminum concentrations greater than about 2×10^{-6} M, the residual turbidity becomes much lower than the minimum (32 nephelometric turbidity units [ntu]) observed in Figure 4-9 at an aluminum concentration of 2×10^{-6} M.

The results plotted in Figure 4-11 were obtained using a silica concentration of 50 mg/L and a sulfate concentration of 3×10^{-4} M. Because the silica concentration is less than the amounts used to obtain the results plotted in Figures 4-9 and 4-10, the amount of aluminum needed for charge neutralization ($\approx 2 \times 10^{-7}$ M) is less than those shown in the previous figures.

Figure 4-11 illustrates an important aspect of hydrolyzing metal coagulants that can complicate the application of EM measurements for dosage control. Note the high residual turbidity between aluminum concentrations of about 10^{-6} and 10^{-5}, suggesting that restabilization has occurred. However, as the aluminum concentration increases beyond 2×10^{-5} M, the residual turbidity decreases to values lower than the minimum observed at the point of charge neutralization (EM = 0). The EM measurement remains at +2 and gives no indication that flocculation efficiency should improve with increasing aluminum concentration beyond the point of charge neutralization.

The results presented in Figure 4-11 can be explained by a consideration of the effect of the aluminum concentration on the magnitude of the product $\alpha_o \varphi$. Apparently under the experimental conditions documented in Figure 4-11 and at aluminum concentrations between 10^{-6} and 10^{-5}, the magnitude of α_o and the product $\alpha_o \varphi$ are not high enough to yield efficient flocculation. However, as the aluminum concentration rises beyond 10^{-5}, the magnitude of φ increases until the modest value of α_o is compensated for and efficient flocculation is obtained.

USE OF EM MEASUREMENTS IN COAGULANT DOSAGE CONTROL

Electrophoretic mobility measurements are used for coagulant dosage control at a number of treatment plants. In one approach, samples of coagulant-treated water are analyzed to determine if coagulant addition is yielding a desired set point EM or ZP. In this type of application, however, the streaming current instrument is likely to give superior results, because it gives a continuous on-line measurement that can be used in a coagulant feed control loop or alarm system.

In another approach, the EM measurement is made using samples of treated suspensions from jar test experiments. This method is perhaps the more appropriate application of manual EM measurements in treatment control. Variables such as solution pH and coagulant dosage can be manipulated in jar test experiments, and the EM measurement can be combined with data for residual turbidity (and possibly other parameters such as residual aluminum) to select effective operating conditions. The city of Durham, N.C., has used this approach for a number of years to effectively control its alum dosage and to select an operating pH for coagulation.

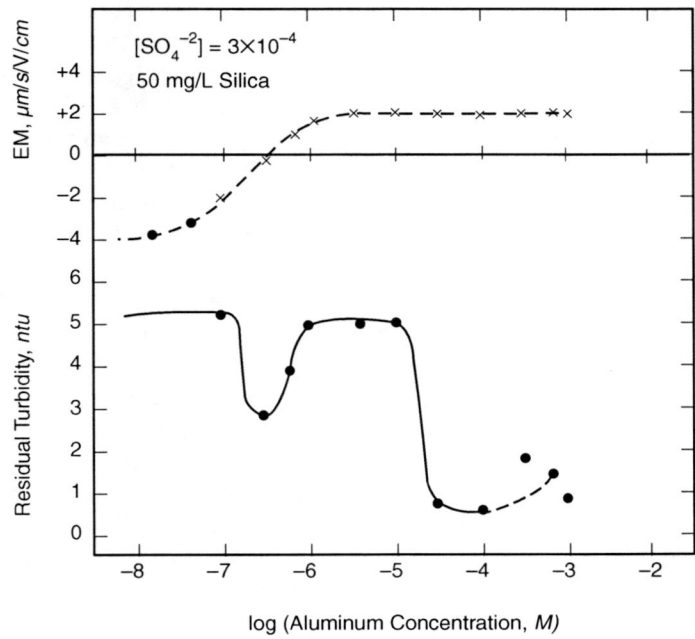

Figure 4-11 Jar test data showing the effect of the aluminum nitrate dosage on electrophoretic mobility and residual turbidity after flocculation and sedimentation for 50 mg/L silica, pH = 6, and $[SO_4^{-2}] = 3 \times 10^{-4}$

The set point EM or ZP in any type of coagulation process control scheme depends on site-specific conditions. For the reasons mentioned previously, no unique set point for EM will give effective particle removal and economical use of coagulant in all situations. Where the EM measurement is based on in-plant evaluations, operators develop and carefully follow set sampling and analysis procedures and then derive empirical relationships between EM values and system performance (e.g., filtered water turbidity) to guide their application of the measurement data. As in any empirical approach, the user of the measurement has to be very careful when conditions fall outside the ranges over which the empirical relationship was developed. In applying the EM measurement in this way, the specifics of the procedure used to make the EM measurements are less important than establishing a set procedure that is consistently followed.

Cleasby et al. (1989) surveyed 23 well-operated filtration plants and reported measurements of ZP for particles in samples from the source water, filter influent, and filter effluent of each plant. The ZP measures for particles from the filter influent were, in most cases, less negative than measures for the particles in the source water and, with several exceptions, were not in the expected 0 to −10 mV range. Cleasby and co-workers concluded that a better sampling procedure might have been to collect the coagulant-treated particles just after the rapid mix rather than after sedimentation. The filter effluent samples yielded poor results, because the researchers had difficulty finding enough particles to track in the electrophoresis apparatus.

SUMMARY

Electrophoresis is an electrokinetic phenomenon in which charged particles in an electrolytic solution migrate under the influence of an applied electric field. The EM of the particles is determined by dividing the measured velocity of migration by the voltage gradient. It is reasonable to assume that the ZP value, which is calculated using the EM, is equal to the diffuse-layer potential of the charged particle.

Theory provides a basis for the assumption that ZP (or EM) is related to the tendency of a suspension to aggregate or deposit. However, theory also provides a basis for the conclusion that the critical value of ZP (i.e., the value that delineates the boundary between a stable and unstable suspension) is a function of variables such as the size of the particles, the ionic strength of the solution, and, in the case of orthokinetic flocculation, the magnitude of the velocity gradient. Thus, constant or carefully controlled conditions are necessary to identfy a "set point" value of the ZP that can be used for coagulant dosage control.

Cationic polyelectrolytes and, under certain conditions, hydrolyzing metal salts affect the destabilization of negatively charged particles by a process that involves adsorption and surface charge neutralization. When destabilization is caused by this mechanism, EM and ZP change from negative to positive at the concentration of coagulant corresponding to maximum instability and, therefore, to maximum particle removal by flocculation/sedimentation or filtration. While this point can be used as a dosage-control set point, it is likely to result in inefficient use of coagulant.

Under some conditions, EM measurements may not provide effective signals to control the dosage of hydrolyzing metal coagulants. These compounds affect the kinetics of flocculation by charge neutralization and by the effect of the voluminous hydrolysis products on the sizes of the particles in the flocculating suspension. As the dosage of the metal salt increases, the controlling effect may result from increasing amounts of metal hydroxide precipitate and their influence on the volume concentration of the suspension; in such a case, EM measurements will not give any indication of the dosage that begins to yield an effective rate of aggregation.

REFERENCES

Cleasby, J.L., et al. 1989. *Design and Operations Guidelines for Optimization of the High-Rate Filtration Process: Plant Survey Results.* Denver, Colo.: American Water Works Association and AWWA Research Foundation.

Hunter, R.J. 1981. *ZP in Colloid Science.* London: Academic Press.

Kim, J.S., and R. Rajagopolan. 1982. A Comprehensive Equation for the Rate of Adsorption of Colloidal Particles and for Stability Ratio. *Colloids and Surfaces*, 4:17–31.

van de Ven, T.M.G., and S.G. Mason. 1977. The Microrheology of Colloidal Dispersions. VII. Orthokinetic Doublet Formation of Spheres. *Colloid and Polymer Science*, 255:468–479.

Ware, B.R. 1977. Applications of Laser Velocimetry in Biology and Medicine. In *Chemical and Biochemical Applications of Lasers*, vol. II. New York: Academic Press (pp. 199–239).

Zeichner, G.R., and W.R. Schowalter. 1977. Use of Trajectory Analysis to Study Stability of Colloidal Dispersions in Flow Fields. *AIChE Journal*, 23:243–254.

This page intentionally blank.

Chapter **5**

Pilot Filters for Process Evaluation and Control

The value of pilot filter testing for process design and control has long been recognized. As early as 1890, Kirkwood observed that "every water has its own special rate of filtration which must be determined by local experiments." From 1895 to 1897, George W. Fuller directed a comprehensive pilot filter testing program at Louisville, Ky., that identified the need for pretreatment and provided the long-standing basis for the design of rapid sand filters at a surface loading rate of 2 gpm/ft^2 (4.9 m/h). Other notable early pilot filter studies include the 1-mgd (0.72×10^6 m^3/year) Chicago demonstration plant directed by John Baylis from 1927 through 1944 and the pilot testing of high-rate filtration conducted by Conley and Pitman in the late 1950s.

From the 1960s through the early 1980s, interest in pilot filter studies remained consistent, confirmed by the many articles in the literature of the period. Recently, new regulatory requirements resulting from passage of the 1986 and 1996 Safe Drinking Water Act (SDWA) Amendments have led to renewed interest in and awareness of microbiological contaminants and the role of filtration in treatment of surface waters and groundwaters influenced by surface waters.

For example, the Surface Water Treatment Rule (SWTR) requires filtration of all surface water supplies and groundwater supplies directly influenced by surface water (although variances are allowed). In addition, the SWTR tightened standards for filter effluent turbidity (0.5 ntu for conventional and direct filtration). Other provisions called for more frequent monitoring of turbidity for new and existing facilities.

Recognition of these new requirements, coupled with increasing costs for treatment facility construction and operation, has led many utilities, regulatory agencies, and consultants to reevaluate traditional filtration plant designs and operation guidelines. Attention is focusing on new technologies, new applications of existing technologies, and alternative methods for process control. The result has been a continuing increase in interest in pilot testing, particularly in the use of pilot filters to evaluate designs, upgrades, and operational control of water facilities. In the design process for a new facility, pilot studies should be conducted for a period sufficient to account for seasonal water quality changes.

PILOT FILTER APPLICATIONS

Pilot filters and studies involving their use can be valuable tools in many areas of water treatment plant design and control. The following sections present brief descriptions of each of these areas.

Evaluation and Optimization of Pretreatment Processes

Ever more stringent turbidity regulations increase the importance of pretreatment (coagulation, flocculation, and sedimentation) for optimization of filter performance. Coagulation and flocculation are the most crucial steps in successful operation of the clarification and filtration processes.

Pilot filter performance can be used to evaluate the suitability of alternative pretreatment schemes and to optimize pretreatment. Jar testing provides a satisfactory method for screening alternative coagulant doses, but it does not adequately model filterability. In addition, jar tests are conducted in batch mode and do not adequately model the effects of hydraulic dynamics on treatment performance. Pilot filters, however, are operated in a continuous-flow mode. If properly designed and operated, they can model full-scale filter response to varying pretreatment conditions. Pilot filter effluent quality and head loss development respond rapidly and readily to changes in pretreatment conditions, such as coagulant dose, point of coagulant addition, coagulant aid addition, rapid mix energy, and flocculation energy. Pilot filters can be used prior to finalizing a design to select pretreatment design criteria, and they can be used at existing plants to help optimize pretreatment performance.

Optimization of New Filter Designs

Pilot filters are excellent tools for evaluating alternative filter designs. Pilot filters can be used to optimize the following filter design criteria:

- Media configuration (rapid sand, dual-media, multimedia, monomedia)
- Media size
- Media depths
- Filtration rate
- Rate control methods
- Required head
- Net production of water of acceptable quality
- Filter backwash rate

Pilot filters can also be used prior to design to evaluate the effectiveness of polymer filter aids, the effect of filter to waste, and other operational features.

Evaluation of Nonconventional Technologies

In general, any consideration of "nonconventional" treatment processes for a new facility design should include studies using pilot filters as an essential predesign task. The state of the art of such processes has not progressed far enough to allow reliable prediction of performance without pilot studies. As a result, design criteria and estimates of construction or operating costs cannot otherwise be developed with confidence.

For a surface water treatment plant with turbidity and color removal as major objectives, the following types of filtration or pretreatment processes should be pilot tested before finalizing a design:

- Dissolved air flotation/filtration

- Direct or in-line filtration

- Shortened flocculation times

- High-rate sedimentation (plate or tube settlers)

- High-rate filtration ($\geq$ 6 gpm/ft^2 [14.6 m/h])

- Prefiltration oxidation/microflocculation

- Contact flocculation/clarification

Pilot testing should also be conducted before design of a new treatment facility for any source water with highly variable characteristics or particularly challenging water quality problems. Source water quality parameters that deserve consideration in pilot filter testing include high algal counts, iron, manganese, tastes and odors, and/or the presence of a wastewater discharge.

Demonstration of Process Applicability and Performance

Demonstration of process applicability and performance is sometimes necessary to give the design engineer and utility confidence in the process and, in many cases, to obtain regulatory agency approval of the design. Many states, such as California, Florida, and Virginia, often require pilot testing before approving designs incorporating nonconventional treatment processes.

Optimization of Process and Operational Control

The nature of the particles delivered to the filters (i.e., size, density, shear resistance, charge) following pretreatment determines the performance of the filtration process. Pilot filters can serve as valuable and relatively simple tools for controlling the coagulation process and maximizing the effectiveness of the filtration process.

In-line pilot filters allow direct measurements of the filterability of coagulated water under actual plant conditions. This application provides a method for alerting operators to changing source water conditions and for controlling coagulation treatment. Rapid mix discharge from a full-scale plant may enter a pilot filter equipped with continuous filtrate turbidity and pH monitors, providing an early detection method to help determine conditions for proper coagulation. Degradation of pilot filtrate quality can rapidly indicate coagulant feed system failure (within about 15 min). As a result, operators can adjust or repair the coagulation process before improperly coagulated water fills the full-scale plant and creates an upset condition that will take several times the plant's hydraulic detention time to rectify.

The Commonwealth of Virginia requires some form of on-line monitoring for filtration plants operated at high rates (more than 2 gpm/ft^2 [4.9 m/h]). This monitoring has typically entailed the use of zeta meters or pilot filters. Recently, streaming current monitors have been installed in several plants for this purpose.

Beyond serving as alarms when improper coagulation occurs, pilot filters can also be used to optimize coagulation processes in existing plants. The effect of adjustments to the full-scale coagulation treatment on plant performance can be rapidly screened and evaluated with pilot filters.

PILOT FILTERS FOR COAGULATION CONTROL

Pilot filter coagulant control systems were originally developed to support training of water treatment plant operators. Such a control system provides operators with an early warning of effluent turbidity problems, which enables them to alter coagulant dose before plant effluent quality begins to degrade.

The first practical and successful coagulant control system using a pilot filter was installed in Eugene, Ore. The system was composed of a 5-in. (125-mm) plastic tube with 30 in. (75 cm) of coal–sand media and a flow-through turbidimeter. This system allowed operators to reduce alum dosages, while maintaining turbidity levels less than 0.15 ntu and increasing filtration rates.

A coagulation control application generally includes two pilot filters, one on-line and the other on standby until the on-line filter requires backwashing. Nonionic polymer is added to prevent premature turbidity breakthrough of a properly coagulated water. The nonionic polymer increases the shear strength of the floc particles, encouraging adherence to one another and to the filter media. This step essentially allows filtration to proceed without time for flocculation and sedimentation. Pilot filters generally require backwashing every 1 to 4 h to maintain available head and sensitivity to coagulation performance.

Pilot filter coagulant control systems provide plant operators with an early warning of improper coagulation conditions, even under rapidly changing source water conditions. One drawback to such a system involves coagulant dose evaluation. As an example, when breakthrough occurs in the system, the operator does not know whether the coagulant dose is too high or too low. In addition, the system must be properly maintained, which adds to the responsibilities of plant operators. The advantages and potential drawbacks of pilot filter coagulant control systems must therefore be weighed and compared to those of other coagulation control systems before a decision is made to install a system.

PILOT FILTER DESIGN

Pilot filter systems can be custom designed to suit unique needs, if vendor-designed equipment is not desired. The following sections discuss some general considerations for preparation of a custom pilot filter design.

Determining the Filter Diameter

The appropriate diameter or cross-sectional area of a pilot filter depends on its intended application. The filter should be sized to provide an appropriate model of a full-scale system. The safest approach is to determine the minimum column diameter that is appropriate for its intended use.

The minimum diameter of a filter column is determined by the size of media to be pilot tested. The cross section of a filter column should measure approximately 50 times the diameter of the largest media to be tested. For example, a pilot filter for testing 2-mm anthracite requires a column diameter approximately 4 in. (100 mm) or larger. For most rapid sand or monomedia designs, a 4-in. (100-mm) diameter column should be considered the minimum, and 6-in. (150-mm) is a more desirable diameter.

Length of a Filter Column

A pilot filter column's length depends on several considerations:

- Media depth
- Underdrain design

- Support media

- Differential head requirements

- Freeboard

Media depth, underdrain design, and freeboard are normally straightforward design considerations. However, the depth required to accommodate a desired differential pressure and the depth of the support media depend on the objectives of the study and any constraints on the pilot filter installation.

Ideally, a pilot filter should be designed to operate by gravity. The filter column must then extend from 6 ft (1.8 m) to over 10 ft (3.0 m) above the top of the media. The overall filter height ranges from 13 ft (4.0 m) to greater than 18 ft (5.5 m). Many pilot filter locations (i.e., filter galleries) will not accommodate this length. As a compromise, pressure columns can be constructed. This option reduces the height provided above the media to a value sufficient to accommodate a backwash operation without washing-out media. A minimum height of 2 times the media depth is normally adequate, assuming 75 percent bed expansion during backwash.

The depth of support media is usually not considered a critical parameter in the design of a pilot filter. Normally, the support media can be eliminated from the filter if the underdrain is designed to prevent loss of the small filter media. In some instances, full support media may be desired (i.e., slow sand filters). In this case, the column height will have to provide for this additional depth.

Finally, when possible, the method of establishing the filter height should include a contingency for modifications. Potential variations could include extra depth for increasing the gravitational driving head or increasing media depth.

Filter Backwash

The filter should be provided with a means of backwashing the media and an option for auxiliary media scour. The normal characteristics of full-scale backwash (i.e., velocity, time, and cleanliness) are not determined easily, if at all, with pilot-scale testing. Consequently, the primary objective of most backwash systems is simply to adequately clean the pilot filter media. Backwash rates well in excess of those normally used in full-scale filters should be provided to allow for complete cleaning of the media.

Auxiliary scour can be accomplished through a surface wash and/or air scour. A surface wash should be provided to assist in initially breaking up the media as the backwash operation begins and in agitating it as the process proceeds. The pilot filter media will tend to lift as a unit at the beginning of the backwash cycle. The design may need to include a surface wash nozzle or even side entry nozzles to break up the media so it is not carried out the filter overflow.

Filter Control

Filter control criteria can include rate of flow, water level, or both. Techniques vary for pressure versus gravity filters.

A pressure filter allows at least two methods for controlling the filtration rate. The rate can be controlled by the feed pump or by a flow control valve with a pressure regulator downstream of the filter. The drawbacks to either of these techniques are floc shear (for all but in-line filter simulation) and pressure variations across the filter.

Flow control in a system with a pump can be achieved by use of a variable-rate, positive-displacement pump (Figure 5-1A). A tubing pump works well for this type of service. If possible, centrifugal pumps should be avoided.

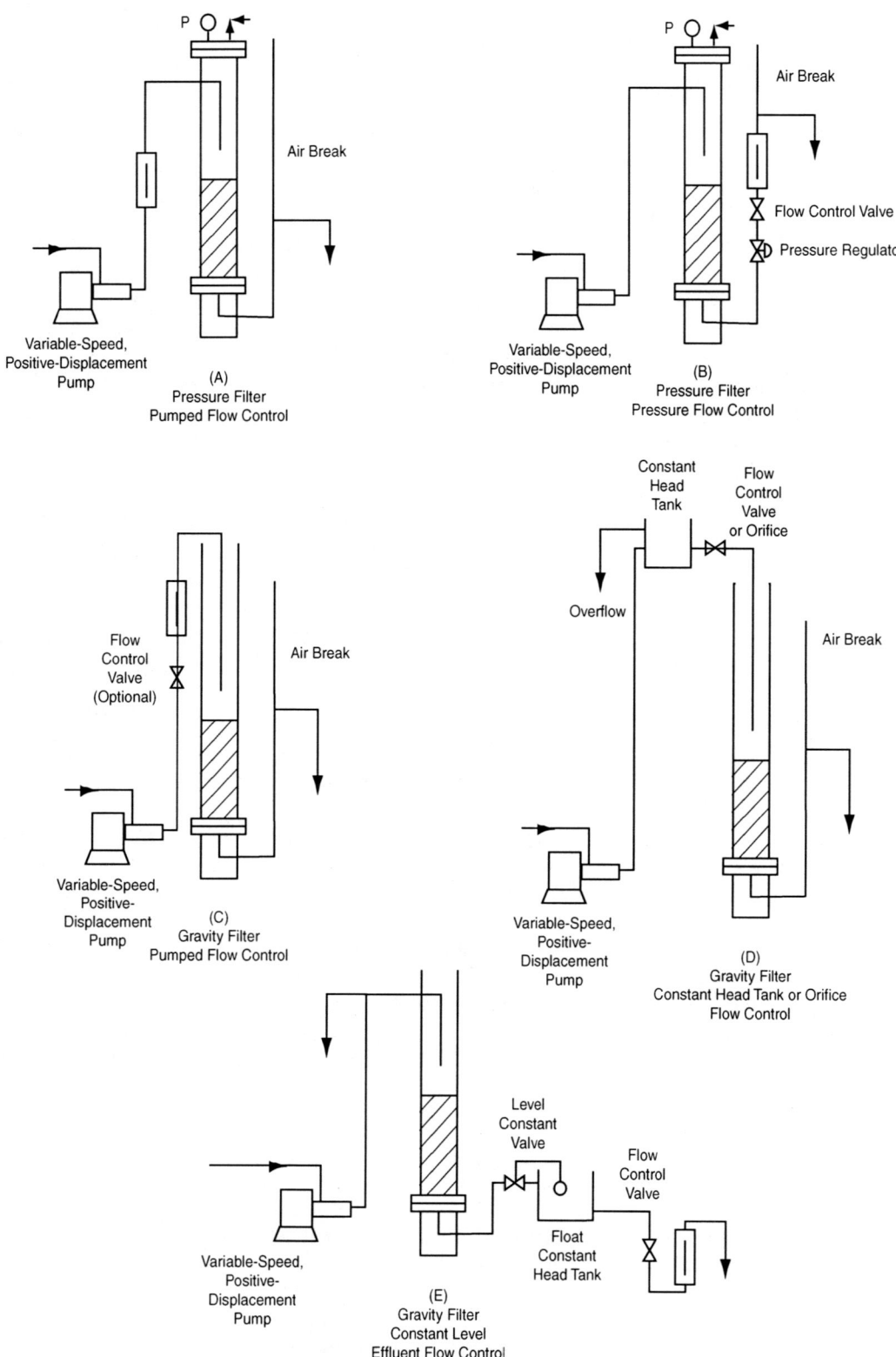

Figure 5-1   Pilot filter control techniques

A pressure regulator and flow control valve (i.e., ball valve) can also be used to regulate flow in a pressure pilot filter column (Figure 5-1B). This method requires a relatively constant influent pressure. The effluent pressure regulator compensates for any pressure increase in the filter bed, and the flow control valve (after the pressure regulator) provides a constant flow given a constant pressure input. This method is not as desirable as the pumped method.

For a gravity-driven pilot filter operation, flow can be controlled by influent pumping, constant head tank and orifice (or flow control valve), or effluent rate control. Filter water level can be controlled with an influent overflow device.

Filter rate control with a pump can be accomplished with a variable-rate, positive-displacement pump similar to that used for pressure filtration (Figure 5-1C). The positive-displacement pump is recommended to limit floc shear, if it is a consideration. A pump can also be used with a flow control valve to regulate the flow. Constant pressure across the control valve is very important to the proper regulation of flow if this method is used.

A constant head tank in combination with an orifice or flow control valve works very well for flow control (Figure 5-1D). The constant head tank will supply the constant pressure required for a consistent flow through the orifice. This configuration works only with a variable water depth type of filter operation.

Use of an effluent rate controller requires the operation of a filter at a constant level (Figure 5-1E). The effluent controller consists of a constant head tank with an influent level control valve (i.e., a float valve) and an effluent flow control valve. The filter effluent into the constant head tank is regulated by the float valve, and the flow rate out of the constant head tank is regulated by the flow control valve.

Other control techniques may work, as well, such as a filter level control valve or an effluent flow control valve linked to effluent flow measurement. However, these methods are not common and the techniques presented in Figure 5-1 can normally be used with less difficulty.

Filter Monitoring

As a minimum, a pilot filter should be equipped with means of measuring flow, influent turbidity, effluent turbidity, and differential pressure. The design to accomplish these functions is straightforward, but it must provide for these monitoring activities without disrupting the flow through or pressure across the filter.

Consideration should also be given to measuring pressure and taking samples at various locations along the length of the filter bed. To allow these capabilities, a pilot filter needs a sample tap that will not significantly change the flow pattern in the filter (i.e., limit flow). A screen is also needed to prevent media from entering and plugging the tap.

All of the measurements can be based on grab (i.e., manual) samples, continuous sampling, and/or continuous recording. Most pilot-plant operators have found that continuous measurement and recording of head loss, influent turbidity, and effluent turbidity provide for significantly improved operations and meaningful results. To limit instrumentation, a configuration might share an instrument among sample sites. This design goal can be accomplished with automatic valving and microprocessor data recording. The cost for including this type of sophistication is not significant when compared to the quality of data collected, improved operations, and limited operator attention.

Materials of Construction

Pilot filters are generally constructed of metal, glass, or plastics. Comparatively large filters (i.e., 1 m square) can be constructed of wood. The most common materials are clear PVC 4-in. (100-mm) columns or clear acrylic 4-in. to 6-in. (100-mm to 150-mm) columns. PVC is easy to work with, while acrylic provides a clearer material through which to view the filtration process.

In some instances, objectives of the pilot testing activities determine the materials to be used. For example, if very low levels of organics are of concern, PVC and/or solvent-welded PVC may not be acceptable choices.

Piping can consist of metal pipe, PVC pipe, and Teflon or plastic tubing. Normally a combination of PVC piping and flexible tubing is used.

Filter Media Loading

Filter media should be loaded by weight. This practice assures that each pilot filter receives the same quantity of media regardless of the differences in compaction that occur during loading. The level (height) of the media after backwash should also be marked as a reference point for future operation and calibration, as the degree of compaction influences porosity of the bed.

Filter Calibration to Existing Filter Operation

Calibration of pilot filters to match existing filter operations will require consideration of both effluent quality and rate of head loss accumulation. Effluent quality calibration is accomplished by varying the treatment system prior to the filter (i.e., the chemicals, rapid mix, flocculation, sedimentation). Once the effluent quality has been calibrated to the existing filters, the head loss characteristics can be adjusted.

Head loss in a pilot filter can be affected by the amount of compaction the media undergoes after backwash. The compaction can be varied by vibrating the filter with a mallet. Calibration consists of varying the amount of compaction (e.g., 1 to 2 in. [2.5 to 5 cm]) after backwash and determining if the rate of head loss buildup in the pilot filter matches that in the full-scale filters. Once the compaction depth is determined, it should be marked on the side of the filter. The media should be returned to this level prior to beginning each filter run.

Data Acquisition

An automatic data acquisition system reduces operator requirements and increases the quality of information that can be derived from the data. At a minimum, data should be collected for:

- Influent turbidity to each pilot filter

- Effluent turbidity from each pilot filter

- Differential head loss across each pilot filter

- Flow from each pilot filter

A number of system components combine to provide a successful means of automating pilot filter data acquisition:

- *Programmable controller*—controls solenoid valves for delivering various sample streams to the turbidimeter

- *Microcomputer*—assimilates electronic signals from the turbidimeter, differential pressure transducer, and flowmeters; displays and stores data; and generates tables and graphic data presentations

- *Turbidimeter*—measures and indicates filter influent and effluent turbidity and sends signal to the microcomputer

- *Differential pressure transducer*—continually measures head loss and sends signals to the microcomputer

Figure 5-2 presents an example of a pilot filter data acquisition system.

OPERATIONAL PROBLEMS AND PITFALLS

A number of problems commonly arise when using pilot filters for water treatment studies or optimization of operations:

- Understaffing
- Insufficient schedule
- Lack of flexibility
- Unrepresentative water quality
- Cheap equipment
- Not anticipating problems
- Inadequate quality control procedures
- Infrequent review of data

Pilot filter operations (studies and full-scale) require good operator attention to produce meaningful and useful results. Even with automated systems to perform continuous monitoring of filtered water turbidity and head loss development, operator attention is still required. Typical problems include:

- Lines plugging
- Maintaining constant flow
- Observing and eliminating mudballs
- Chemical feed pump failure or deviation from desired rate
- Running out of chemical
- Recognizing and compensating for changes in operations

PILOT FILTER PROCUREMENT AND COSTS

Pilot filters can be obtained in a number of ways. Filters can be constructed by the user (e.g., utility staff), a consultant can be hired to design and/or construct the pilot filters, or a vendor can be contacted to supply filters.

For a short-term study, researchers may prefer to rent equipment or acquire the full operational services of a consultant. A longer study, especially for operational control and routine testing in an existing plant, calls for careful design or purchase and installation of needed equipment.

The basic materials for a filter can cost from a few hundred dollars for simple PVC construction to several thousand dollars for glass or specialized acrylic materials. The cost of the filters is usually minor relative to the cost of the associated monitoring and recording equipment and labor associated with the intended use.

Remember, however, that results from a pilot filter will be only as good as the equipment and personnel involved in its construction and operation. Using low-quality equipment for the assembly of the pilot filters (and associated treatment facilities) will normally result in excessive operating costs or complete failure of the pilot plant. It does not take many failures to warrant the purchase of a reliable pump.

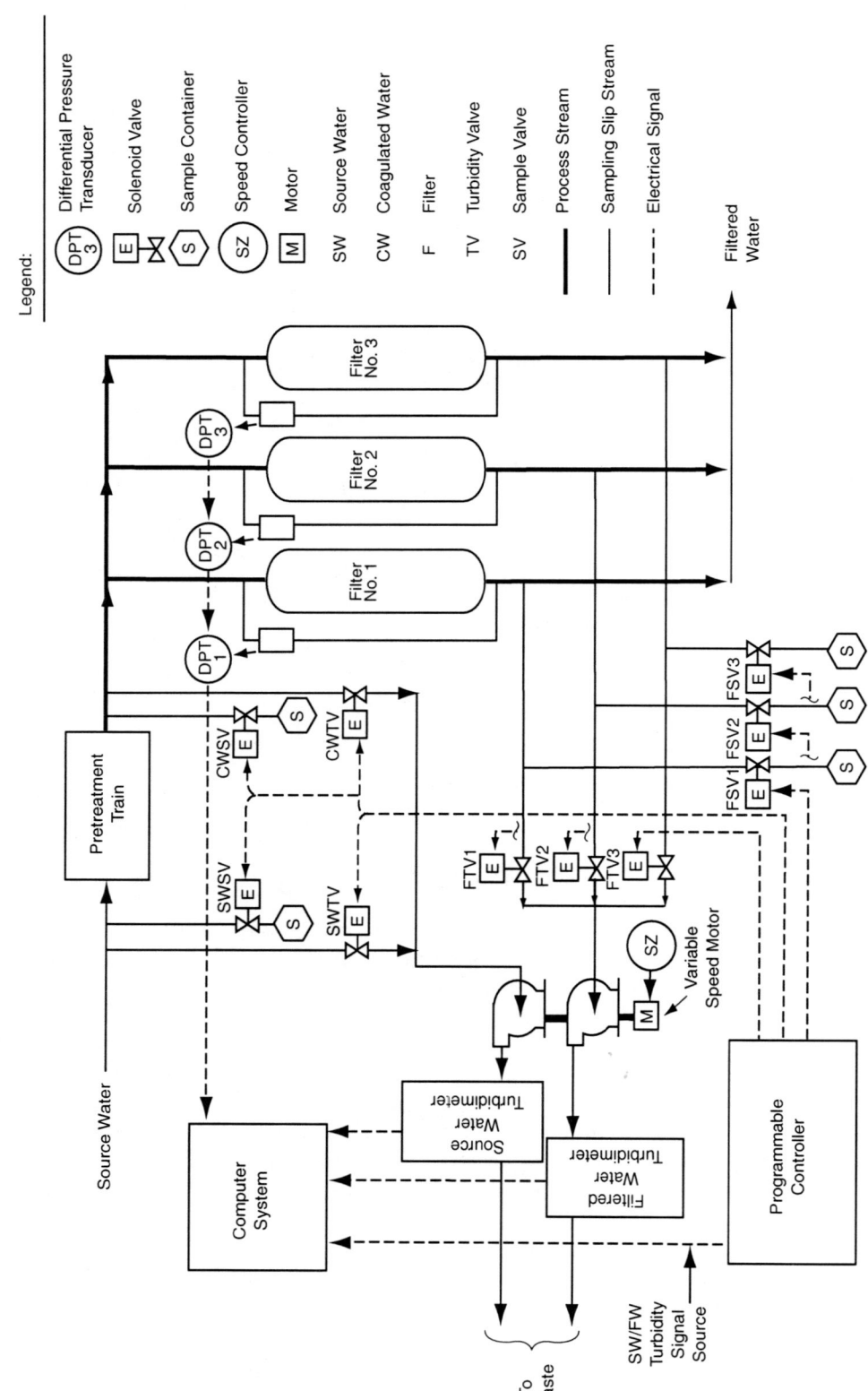

Courtesy of CH2M HILL.

Figure 5-2   Pilot filter data acquisition and control diagram

SUMMARY

Pilot filters provide operating data applicable to many areas of water treatment plant design. In addition, process and operational control can be optimized by the use of pilot filter systems. The design of a pilot filter system depends on its intended purpose. Often the level of quality in design and construction of the pilot filter system determines the quality of the data and information generated by the system. The data and information obtained from pilot filters and studies can help to prevent costly design errors and provide early indicators of problems within an operating plant.

This page intentionally blank.

ADDITIONAL SOURCES OF INFORMATION

Electrophoretic Mobility Measurements

Letterman, R.D. 1991. *Filtration Strategies to Meet the Surface Water Treatment Rule.* Denver, Colo.: American Water Works Association.

Particle Counters

Amirtharajah, A. 1978. Design of Flocculation Systems. In *Water Treatment Plant Design*, R.L. Sanks, ed. Ann Arbor, Mich.: Ann Arbor Science Publishers.

Hannah, S.A., J.M. Cohen, and G.G. Roebeck. 1967. Measurement of Floc Strength by Particle Counting. *Jour. AWWA*, 69:7:843.

Kukshtel, N.S. 1981. An Investigation of Direct Filtration as a Pretreatment Process for Desalination. Unpublished Master's thesis, University of North Carolina, Chapel Hill, N.C.

Lawler, D.F. 1979. A Particle Approach to the Thickening Process. Unpublished Ph.D. dissertation, University of North Carolina, Chapel Hill, N.C.

Monscvitz, J., and D.J. Rexing. 1983. Direct Filtration Control by Particle Counting. *Proc. AWWA Ann. Conf.*, Las Vegas, Nev. Denver, Colo.: American Water Works Association.

O'Melia, C.R. 1972. Coagulation and Flocculation. In *Physicochemical Processes for Water Quality Control.* W.J. Weber, Jr., ed. New York: Wiley-Interscience.

Ramaley, B.L., et al. 1981. Integral Analysis of Water Plant Performance. *Jour. Envir. Engrg. Div., Proc. ASCE*, 107:EE3.

Scarpino, P. 1984. *Effect of Particulates on Disinfection of Enteroviruses in Water by Chloramines.* NTIS, PB 84-190693.

Spectrex Corporation. 1987. Instruction Manual for SPC-510. Redwood City, Calif.: Spectrex Corp.

Tekippe, R.J., and R.K. Ham. 1970. Apparatus to Examine Floc-Forming Processes. *Jour. AWWA*, 62:9:260.

Treweek, G.P., and J.J. Morgan. 1977. Size Distribution of Flocculated Particles: Application of Electronic Particle Counter. *Envir. Science & Technol.*, 11:7:707.

Trussell, R.R., and C. Tate. 1979. Measurement of Particle Size Distribution in Water Treatment. *Proc. AWWA Water Quality Technology Conf.*, Philadelphia, Pa. Denver, Colo.: American Water Works Association.

USEPA. 1987. Draft Guidance Manual for Compliance With the Filtration and Disinfection Requirements for Public Systems Using Surface Water Sources. Washington, D.C.: USEPA, Science and Technology Branch, Office of Drinking Water.

Streaming Current Detectors

Bryant, R.L. 1985. Precise Coagulant Control Using Streaming Current Measurement. *Waterworld News*, 1:3:18.

Veal, C.R., Jr. 1988. Streaming Current Monitor Controls Coagulation. *Opflow*, 14:8:4.

This page intentionally blank.

Index

NOTE: *f.* indicates figure; *t.* indicates table.

AWWA Manuals

M1, *Principles of Water Rates, Fees, and Charges,* Fifth Edition, 2000, #30001PA

M2, Instrumentation and Control, Third Edition, 2001, #30002PA

M3, *Safety Practices for Water Utilities,* Sixth Edition, 2002, #30003PA

M4, *Water Fluoridation Principles and Practices,* Fifth Edition, 2004, #30004PA

M5, *Water Utility Management Practices,* First Edition, 1980, #30005PA

M6, *Water Meters-Selection, Installation, Testing, and Maintenance,* Fourth Edition, 1999, #30006PA

M7, *Problem Organisms in Water: Identification and Treatment,* Third Edition, 2004, #30007PA

M9, *Concrete Pressure Pipe,* Second Edition, 1995, #30009PA

M11, *Steel Pipe-A Guide for Design and Installation,* Fifth Edition, 2004, #30011PA

M12, *Simplified Procedures for Water Examination,* Third Edition, 2002, #30012PA

M14, *Recommended Practice for Backflow Prevention and Cross-Connection Control,* Third Edition, 2003, #30014PA

M17, *Installation, Field Testing, and Maintenance of Fire Hydrants,* Third Edition, 1989, #30017PA

M19, *Emergency Planning for Water Utility Management,* Fouth Edition, 2001, #30019PA

M21, *Groundwater,* Third Edition, 2003, #30021PA

M22, *Sizing Water Service Lines and Meters,* Second Edition, 2004, #30022PA

M23, *PVC Pipe-Design and Installation,* Second Edition, 2002, #30023PA

M24, *Dual Water Systems,* Second Edition, 1994, #30024PA

M25, *Flexible-Membrane Covers and Linings for Potable-Water Reservoirs,* Third Edition, 2000, #30025PA

M27, *External Corrosion Introduction to Chemistry and Control,* Second Edition, 2004, #30027PA

M28, *Rehabilitation of Water Mains,* Second Edition, 2001, #30028PA

M29, *Water Utility Capital Financing,* Second Edition, 1998, #30029PA

M30, *Precoat Filtration,* Second Edition, 1995, #30030PA

M31, *Distribution System Requirements for Fire Protection,* Third Edition, 1998, #30031PA

M32, *Distribution Network Analysis for Water Utilities,* Second Edition, 2005, #30032PA

M33, *Flowmeters in Water Supply,* First Edition, 1989, #30033PA

M36, *Water Audits and Leak Detection,* Second Edition, 1999, #30036PA

M37, *Operational Control of Coagulation and Filtration Processes,* Second Edition, 2000, #30037PA

M38, *Electrodialysis and Electrodialysis Reversal,* First Edition, 1995, #30038PA

M41, *Ductile-Iron Pipe and Fittings,* Second Edition, 2003, #30041PA

M42, *Steel Water-Storage Tanks,* First Edition, 1998, #30042PA

M44, *Distribution Valves: Selection, Installation, Field Testing, and Maintenance,* First Edition, 1996, #30044PA

M45, *Fiberglass Pipe Design,* First Edition, 1996, #30045PA

M46, *Reverse Osmosis and Nanofiltration,* First Edition, 1999, #30046PA

M47, *Construction Contract Administration,* First Edition, 1996, #30047PA

M48, *Waterborne Pathogens,* First Edition, 1999, #30048PA

M49, *Butterfly Valves: Torque, Head Loss, and Cavitation Analysis* First Edition, 2001, #30049PA

M50, *Water Resources Planning,* First Edition, 2001, #30050PA

M51, *Air-release, Air/Vacuum and Combination Air Valves,* First Edition, 2001, #30051PA

M54, Developing Rates for Small Systems, First Edition, 2004, #30054PA

To order any of these manuals or other AWWA publications, call the Bookstore toll-free at 1-(800)-926-7337.

This page intentionally blank.

Printed in the United States
69294LVS00004B/104

9 781583 210550